T0318600

Practical Design and Application of Model Predictive Control

Practical Design and Application of Model Predictive Control

MPC for MATLAB® and Simulink® Users

Nassim Khaled
Controls and Engineering Systems Manager,
Dover, Atlanta, Georgia

Bibin Pattel
Solution Architect, KPIT Infosystems Inc, Columbus, Indiana, USA

Butterworth-Heinemann
An imprint of Elsevier

Butterworth-Heinemann is an imprint of Elsevier
The Boulevard, Langford Lane, Kidlington, Oxford OX5 1GB, United Kingdom
50 Hampshire Street, 5th Floor, Cambridge, MA 02139, United States

Notices
Knowledge and best practice in this field are constantly changing. As new research and experience broaden
our understanding, changes in research methods, professional practices, or medical treatment may become
necessary.

Practitioners and researchers must always rely on their own experience and knowledge in evaluating and
using any information, methods, compounds, or experiments described herein. In using such information or
methods they should be mindful of their own safety and the safety of others, including parties for whom
they have a professional responsibility.

To the fullest extent of the law, neither the Publisher nor the authors, contributors, or editors, assume any
liability for any injury and/or damage to persons or property as a matter of products liability, negligence or
otherwise, or from any use or operation of any methods, products, instructions, or ideas contained in the
material herein.

British Library Cataloguing-in-Publication Data
A catalogue record for this book is available from the British Library

Library of Congress Cataloging-in-Publication Data
A catalog record for this book is available from the Library of Congress

ISBN: 978-0-12-813918-9

> For Information on all Butterworth-Heinemann publications visit our
> website at https://www.elsevier.com/books-and-journals

Working together
to grow libraries in
developing countries

www.elsevier.com • www.bookaid.org

Publisher: Mara Conner
Acquisition Editor: Sonnini R. Yura
Editorial Project Manager: Lindsay Lawrence
Production Project Manager: Vijayaraj Purushothaman
Cover Designer: Mark Rogers

Typeset by MPS Limited, Chennai, India

Dedication

I dedicate this book to my beloved parents. They are good peasants who tried their best to raise a good peasant. They encouraged me to pursue my education. I am indebted to them for life.

—Nassim

I dedicate this book to my dear parents, who did not have the fortunate circumstances for getting good education, but understood the value of it and worked really hard to get quality education for myself and my brothers. I would like to also dedicate this to my dear wife and daughter—they always have been my inspiration and hope. I would like to acknowledge my brothers who encourage me all the time. Thank you all for your support and being there for me whenever I needed.

—Bibin

Contents

Preface

One of the main hurdles prohibiting the use of model-based controllers (such as Model Predictive Control) in various industries is the lack of familiarity with the technique in addition to cost. For every hundred control engineers we've worked with, less than five actually get to design and tune a controller (usually PID). They are either working on the device drivers, maintaining or updating a legacy code, running testing and validation, or performing data analysis. The few who get to design the controller are usually updating a previous design or adding minor features. Only the lucky ones are actually brainstorming robust control techniques. We are from the lucky few who were privileged to do brainstorming on robust control techniques in industrial applications.

We believe in the value of simulation models that mimic the laws of physics. We are promoters of robust control techniques. Model Predictive Control (MPC), with its various flavors, is a great model-based control technique. In this book, we propose a streamlined approach for designing MPC. We hope that more industrial companies will embrace this practical and fascinating technology.

Acknowledgments

For professionals like us who have full time 8−6 jobs and a family to take care of, publishing a book is not an easy task. We were fortunate over the past 4 years to have the support of several parties. They were instrumental in bringing this material to production. We would like to take this opportunity to mention the main contributors and thank them for their support.

The MathWorks sponsored this book under their Book Program. They provided us with MATLAB and Simulink licenses for the past 3 years. In particular, we would like to thank Ms. Fernandes for her continuous support and providing sponsored licenses.

We would also like to thank Dr. Sharif Aljoaba for coauthoring Chapter 8, MPC Design for Photovoltaic Cells, and providing the simulation model for the photovoltaic cell. It served as a good application for MPC.

Elsevier provided us with a smooth process, from review to production. In particular, Ms. Yura and Ms. Lawrence were prompt in responding to our enquiries.

We also extend a special thanks to SAE who allowed us to republish our work on MPC for air-handling controls of diesel engines.

Chapter 1

Introducing the Book

1.1 INTRODUCING THE AUTHORS

One of the unique features of this book is the fact that neither one of the authors learned Model Predictive Control (MPC) in a classroom setting. They both had to self-learn the theory, design, and implementation of MPC in an application-oriented fashion. They have gone through the pain of failed designs and tunings in their industrial experiences. They have learned coding tricks, automated multiple MPC design techniques as well as robustness best practices that they wanted to share with the industrial and academic world. This important fact allows the reader to understand the how this book came about and what benefits he/she will reap from reading this publication.

Both authors are academic and industrial controls experts. In addition to MPC, they have studied, designed, and implemented several controller and observer strategies such as sliding mode, fuzzy logic, adaptive techniques, linear and nonlinear PID. Many of their designs were implemented and integrated into industrial products such as diesel engine control, onboard diagnostics, automated testing stands, retail, and industrial refrigeration.

The authors have created a webpage that has additional resources related to the book. The reader can contact the authors with questions, feedback, seminars, or consultancy inquiries. As the authors receive a lot of similar requests, please expect some delay in response. The objective of the authors is to connect with readers, maximize the benefit to readers, as well as improve the quality of the material related to this book.

www.practicalmpc.com

1.2 PRACTICAL APPROACH TO MPC

Since the eighties, a significant body of books have described theory in addition to examples of Model Predictive Control (MPC). Academics were drawn to MPC since it provides a streamlined solution for solving Multi-Input Multi-Output control problems that are subject to constraints in the system. Furthermore, MPC provides the designer with the ability to handle the instantaneous as well as future performance of dynamic systems. In the case of industrial process control, the Honeywell industrial MPC controller [1] was designed to handle complex industrial process control that can't be handled

Practical Design and Application of Model Predictive Control.
DOI: https://doi.org/10.1016/B978-0-12-813918-9.00001-0
© 2018 Elsevier Inc. All rights reserved.

with the traditional and popular PID. Yet, it seems that the popularity of MPC hasn't gained much traction in many industries, such as the automotive world. It is rarely cited that MPC solutions made their way into production electronic modules for vehicles. The authors believe that this is primarily due to the significant resources which are required to change existing procedures in software development by switching to MPC, limited capability of automotive electronic control units (ECUs) in terms of throughput and memory, as well as the lack of automotive control engineers who are well versed in MPC. Moreover vehicle manufacturers are still finding ways to design their closed loop controllers without using MPC. Nontechnical budget holders in the automotive world continuously pose questions such as: Why should we change the controller if it works? Why do we need to invest in new procedures to adopt MPC? Do customers care about having an MPC controller in their vehicles instead of nested PID loops? These are all valid questions and the challenge that technical leaders face is how to quantify control robustness as cost savings. The authors believe that until the complexity level of designing and tuning control software for automotive applications in particular, and other industries in general, reaches unmanageable levels through traditional control techniques, there will not be wide adoption of MPC in the industry.

In an attempt to understand the academic interest in MPC compared to other traditional control techniques, the authors used *books.google.com/ngrams* which scans a serious volume of books written in English. The authors searched for the frequency of usage of the following case-insensitive keywords: Model Predictive Control, PID Control, Sliding Mode Control, State Feedback Control. Fig. 1.1 shows the search results from 1970 till 2008 no data was available after 2008. At the beginning of the millennia, the frequency of usage of MPC surpassed PID as well as other control techniques. Fig. 1.1 is a good indication that there is an increasing academic interest in MPC, especially with the increased interest in the internet of things (IoT) and smart devices.

The theme of this book is streamlining the design, tuning, and deployment process of linear MPC. This will allow a wider spread of a very

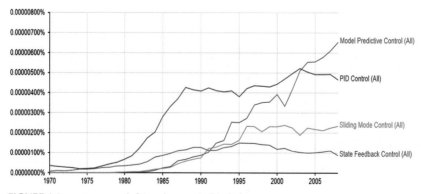

FIGURE 1.1 *ngrams* search for various control techniques.

capable control strategy that the authors believe will be an essential part of the technology revolution of IoT, smart devices, and digital twins. The methodology the authors use to educate readers is through solving real world applications. The control problems discussed in this book are challenging and nonlinear. The authors spent a significant amount of time modeling the dynamics of the presented problems as well as designing and tuning the MPC controllers. Where possible, all the challenges the authors encountered were documented so that readers can benefit from the lessons learned. The challenges ranged from the system identification of the plant, design of MPC in MATLAB and Simulink, the untold tuning art of MPC as well as simulating the MPC with the nonlinear plant in Simulink. Except for Chapter 10, all the plant models and the designed MPCs can be downloaded from the book's website. The authors believe in open-sources sharing to advance science and promote model-based control approaches.

To enrich the expertise pool that contributed to the book, the authors reached out to Dr. Sharif Aljoaba to leverage his experience in modeling and control of photovoltaic cells.

All the examples provided in this book have been developed using MATLAB R2017a. The toolboxes used are: Model Predictive Control and System Identification. The operating system used is Windows 10, 64 bit. If the reader doesn't have MATLAB, he/she can contact Mathworks for a trial version.

1.3   ORGANIZATION OF THE BOOK

To the extent possible, the authors made the chapters independent from one another. However, the book was written with an increasing level of complexity. The process of designing, tuning, and implementing MPC is re-iterated in all the chapters. Chapter 2 briefly discusses the theoretical foundation of MPC which will set the stage for the subsequent implementation of MPC. The chapter suggests a hypothetical PID controller that resembles MPC which is used to familiarize the reader with the concept of MPC. The second half of the chapter introduces MPC. Chapter 3 covers a streamlined approach for system identification and MPC design. The approach is applied on a double mass-spring system. In Chapter 4, a nonlinear model of long ship navigating at sea is introduced. The model is used as a testbed for linear system identification. The same ship model is used to implement and test MPC in Chapter 5. In Chapter 6, the concept of multiple MPC will be introduced. The controller is applied to the ship model of Chapter 5, but the space of operation is expanded which necessitates the use of multiple linear MPC controllers. Parallel Computing Toolbox is demonstrated to design MPC. Additionally, the challenges with frequent switching among the modes is tackled. A hysteresis logic is implemented to mitigate actuators' chattering.

In Chapter 7, the robustness of the multiple MPC designed in the previous chapter is challenged through Monte-Carlo simulations. In Chapter 8, the novel model for photovoltaic cells developed by Dr. Sharif Aljoaba is introduced and used as a testbed for the design of MPC. Chapter 9 describes the process of embedding MPC in a real-time target application. Arduino Mega is used to test the developed MPC controller. The book is concluded with Chapter 10 which demonstrates a real application of MPC for the control of a complex air-handling diesel engine. Simulation as well as experimental results are shown.

1.4 SOFTWARE AND HARDWARE REQUIREMENTS

The MATLAB version that was used to develop the codes is R2017a.

MATLAB Toolboxes that were used are: MATLAB, Simulink, Model Predictive Control Toolbox, System Identification Toolbox. In addition to these toolboxes, MATLAB Parallel Computing Toolbox was used in Chapter 6, and in Chapter 9 we used Embedded Coder, MATLAB Coder, and Simulink Coder.

The operating system is Microsoft Windows 10 Home Version 10.0 (64-bit).

1.5 DOWNLOADING THE SOURCE CODES

The authors are big advocates of open-source as means to share and spread scientific practices. All the codes are available on Mathworks' website for free download. Please reference the book in case you use the codes in your publications.

To download the codes, follow the below link (file exchange) and search for the ISBN or title of the book.

https://www.mathworks.com/matlabcentral/fileexchange/

Additionally, the hardware setup for Chapter 9 and other application problems can be purchased with the MPC controller loaded on Arduino Mega from the book's website:

https://www.practicalmpc.com/mpc-store

REFERENCE

[1] http://www.automationworld.com/process-control-software/pulppaper-and-power-gen-industries-embrace-advanced-control

FURTHER READING

http://discover.rockwellautomation.com/Media/Files/Chlorine-DioxideAppProfile.pdf

Chapter 2

Theoretical Foundation of MPC

2.1 INTRODUCTION

This chapter introduces readers to the fundamentals of linear MPC. Intentionally light on formulation, this chapter serves as a first encounter with MPC. In section 2.2, we provide rough guidelines to allow the reader to choose between PID and MPC control strategies. In section 2.3, we build a modified version of PID controller that resembles MPC. The proposed PID controller has a prediction horizon. In section 2.4, the control horizon concept is introduced and integrated with the suggested PID controller. We introduce MPC in section 2.5. The optimization solver for MPC is introduced in section 2.6. Section 2.7 is dedicated to introducing the MATLAB Model Predictive Control toolbox. The chapter is concluded with a listing of references.

2.2 PID OR MPC

Conventional control techniques, such as PID, rely on current and previous measurements to regulate systems. PID does not (explicitly) use the dynamic characteristics of the system to command the manipulated variables. Optimal performance of the closed loop system is achieved through tuning the proportional, integral, and derivative terms.

Additionally, PID cannot naturally handle known constraints of the system such as maximum speed of a vehicle in the case of cruise control.

Based on the authors' industrial experience, PID is still the first choice for industrial feedback control mainly due to the ease of understanding its operation, ease of tuning and its widespread use in the industry. Control engineers are using creative techniques to handle the limitations of PID. These include, but are not limited to, constraints, multi-input multi-output control in addition to time-varying dynamic systems such as aging impact on the dynamics of a system.

A question that is often asked is when control engineers should use MPC versus PID? The flow chart below helps the engineer decide when is the usage of MPC warranted in the design. The flowchart in Fig. 2.1 should serve as a guideline, however, the control problem in hand needs to be

Practical Design and Application of Model Predictive Control.
DOI: https://doi.org/10.1016/B978-0-12-813918-9.00002-2

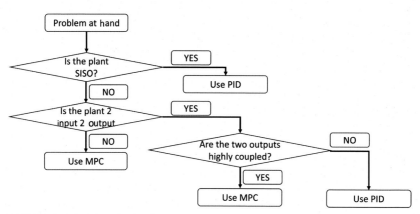

FIGURE 2.1 Flowchart to select PID vs MPC.

studied on a case by case basis. The referenced PID can be a single PID, staged PID, or a variant of PID.

2.3 HYPOTHETICAL PID WITH A PREDICTION HORIZON

A MPC controller relies on a plant model of the physical system to simulate (optimize) various responses of the system, and choose the best (optimal) path. In this section, we propose a hypothetical controller that is based on PID. The controller has common elements with MPC. It would serve as a stepping stone to understanding the fundamentals of MPC in the subsequent sections.

As an illustration, we will use the cruise control of a vehicle to explain the methodology of a hypothetical PID controller. The input is force whereas the output is vehicle speed (Fig. 2.2).

Fig. 2.3 shows the block diagram of a traditional discrete PID controller implemented to control the vehicle's speed.

The proportional (P), integral (I), and derivative (D) parameters are the only tuning parameters used in PID. Eq. (2.1) shows the discrete form of PID

$$u(n + 1) = P \times e(n) + I \times \sum_{i=1}^{n} e(i) + D\Big(e(n) - e(n - 1)\Big) \qquad (2.1)$$

where

u is the manipulated variable
e is the error, which is the difference between the reference and the measured output
n is index representing the current sample in time

Fig. 2.4 shows the results of simulating the closed loop system for the vehicle with a discrete PID controller executing at a sampling rate of 0.5 seconds.

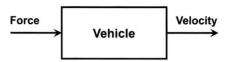

FIGURE 2.2 Vehicle model.

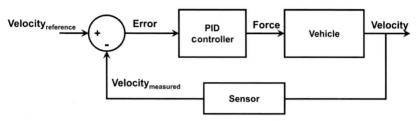

FIGURE 2.3 PID controller.

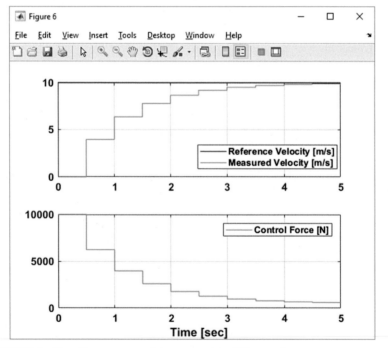

FIGURE 2.4 Closed-loop response of the system with a PID controller.

Let's try various tunings to get better performance from the system. Without loss of generality, we will try three PID tunings and compare the tracking.

Tuning 1 seems to yield the best tracking. Tuning 3 has overshoot which is highly undesirable for speed control.

Let's create a hypothetical controller that can:

1. internally simulate the three tunings highlighted in Fig. 2.5
2. compute the tracking error
3. choose the tuning that has minimal tracking error

Based on the above hypothetical controller, Tuning 1 will be chosen after simulating the three tunings for 5 seconds (or 10 samples). The 5 seconds window (or 10 samples) is referred to as the prediction horizon, p.

We will define two spaces to clarify how the hypothetical controller executes:

— controller simulator
— controller executer

In the controller simulator space, the above three tunings are simulated at time $t = 0$ having a plant model of the system and the PID running in closed-loop. The controller simulator is concluded in less than 1 step size (0.5 seconds).

In the controller executer, the best tuning is used from the first sample time and moving forward in time (in this case 5 seconds).

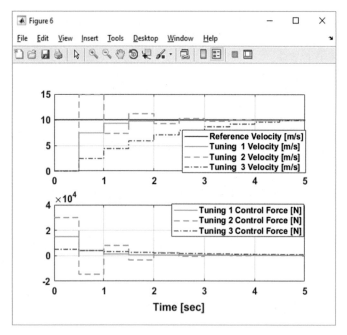

FIGURE 2.5 Tracking performance for three PID tunings.

One of the main issues with implementing such a hypothetical controller is the fact that the plant model used to simulate the performance of the tunings generally doesn't match the actual dynamics of the plant, it is usually a good approximation, but not an exact match. Furthermore, measurement noise and inaccuracies coupled with unmeasured factors (e.g., air force in the case of the vehicle) will add to the inaccuracy of the plant model used in the controller simulator.

A good approach to remedy this problem is to repeat the controller simulations at each time step. Available measurements can be used to "correct" the plant inaccuracies. The optimal tuning is then implemented at the next time step. Fig. 2.6 shows the hypothetical PID controller with prediction horizon capability. The controller also has the correction of the internal states of the vehicle model using measurements.

In the controller simulator space, the three tunings are simulated at time $t = 0$ having a plant model of the system and the PID running in closed-loop. The controller simulator is concluded in less than 1 step size (0.5 seconds) for a prediction horizon of 5 seconds (or 10 samples). Unlike Fig. 2.5, in the controller executer, the best tuning is used from the first sample time (0.5 seconds) and held constant for the second sample time (1 second). Note that at time zero seconds, the controller has to assume an initial value.

A different way to think about the concept of updating the simulation model is to use an analogy with the way you plan your future. Say that you are accustomed to planning your future for the next five years. You are effectively simulating the outcome and assuming a certain model for the dynamics of your life. When you get feedback from the real world such as unplanned opportunities, birth or death in the family, then you re-plan for

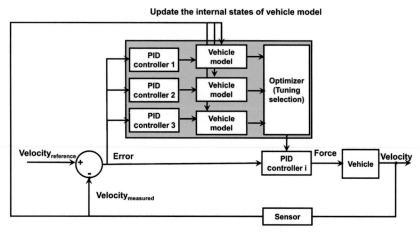

FIGURE 2.6 Hypothetical PID control with Prediction Horizon.

the next five years you are effectively simulating the dynamics of your life while updating the new states of your model.

2.4 HYPOTHETICAL PID WITH A PREDICTION AND CONTROL HORIZON

In the previous section, a hypothetical PID controller was proposed which has a prediction horizon of 10 samples. We mentioned that the controller needs to simulate the three tunings for the duration of the prediction horizon. The controller must execute the simulation, and run the optimization before the lapse of the time step of 0.5 seconds. When you have a very capable microprocessor, this is not an issue, especially for a simple example like the vehicle speed control that we introduced. If the microprocessor is not capable (in many cases it is not capable to run MPC refer to Chapter 9, Real Time Embedded Target Application of MPC and Chapter 10, MPC Design for Air-Handling Control of a Diesel Engine for some details), we need to use other techniques to reduce computation time. One of these techniques is to reduce the *control horizon, m* (definition to follow). The control horizon is the number of allowed moves for the duration of the simulation phase. In section 2.3, we assumed that the PID controller value can change 10 times (refer to the Control simulator portion of Fig. 2.7). The prediction and control horizons were equal. In most cases, the control horizon, m, is chosen to be less than p. The value of the control command is held constant from simulation step size of $m + 1$ to p.

Fig. 2.8 illustrates the controller simulation with prediction horizon of $p = 10$ and control horizon of $m = 6$. The minimum tracking error was that of PID Tuning 1, and thus in the controller execution the commanded value for the first step size is that of PID Tuning 1. This result matches the one in Fig. 2.7 where $p = 10$ and $m = 10$ despite the reduction in the control horizon. The vertical line in the middle plot of Fig. 2.7 marks the end of the control horizon after which the command from PID is kept constant.

Fig. 2.9 illustrates the controller simulation with prediction horizon of $p = 10$ and control horizon of $m = 2$. The minimum tracking error was that of PID Tuning 3, and thus in the controller execution the commanded value for the first step size is that of PID Tuning 3. When the control horizon was significantly reduced, there was a significant deterioration in the tracking error in the controller simulation.

Since we are showing only three PID scenarios, the impact of reducing the control horizon was exaggerated. An MPC optimizer goes through more than three iterations to find the optimal solution. Furthermore, the optimization is repeated at each time step. These two facts allow the reader to get acceptable results even with a control horizon of 20% of that of the prediction horizon. Tuning the control horizon, as will be seen in the subsequent chapters, is problem-dependent, but we advise the reader to increase it if

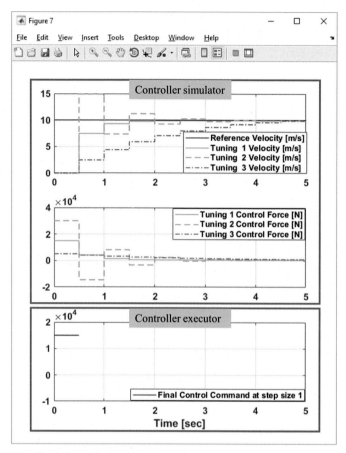

FIGURE 2.7 Simulation of the hypothetical PID control with prediction horizon.

there is no computational throughput problem. Otherwise, having a control horizon that is equal to 20%−60% of the prediction horizon can be used.

2.5 INTRODUCTION TO MPC

Model Predictive Control, as the name suggests, is a feedback control technique that relies on a model. It also relies on a real-time optimization solver. Other names for MPC are receding horizon control and quadratic programming control. We can also call MPC a new name which is Optimization and Prediction Control (OPC). It is a made-up name by the authors to help the reader remember that MPC also carries an optimization element. Fig. 2.10 shows a simplified block diagram of MPC.

In the previous section, we said that the hypothetical PID controller has an optimizer (Fig. 2.6) that computes the minimal tracking error. In MPC,

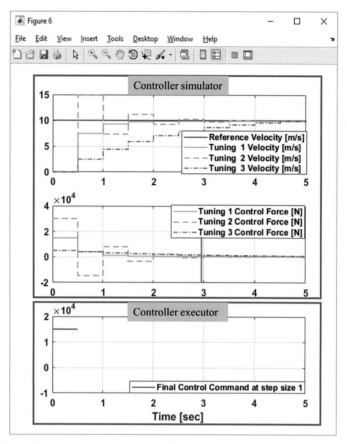

FIGURE 2.8 Simulation of the hypothetical PID control with $p = 10$ and $m = 6$.

the tracking error is one part of the cost function. The cost function of MPC is made of four elements:

$$J(z_k) = J_y(z_k) + J_u(z_k) + J_{\Delta u}(z_k) + J_\varepsilon(z_k) \qquad (2.2)$$

where:

z_k is the sequence of manipulated variables from sample k to $k + p - 1$.
J_y is the cost function for output reference tracking (or tracking error)
J_u is the cost function for manipulated variable tracking (or deviation from nominal manipulated variable)
$J_{\Delta y}$ is the cost function for change in manipulated variables
J_ε is the cost function for constraint violations

Fig. 2.11 shows a detailed block diagram of MPC as implemented in MPC Toolbox documentation. The dashed-line block Cost Function shows the four elements of Eq. (2.2). It is worthwhile noting that MPC doesn't react

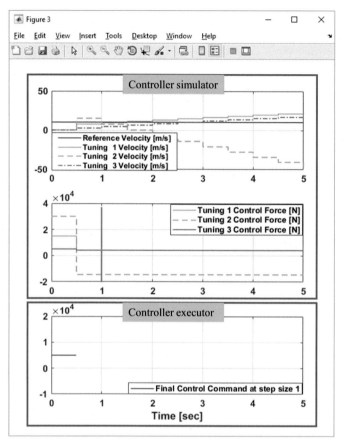

FIGURE 2.9   Simulation of the hypothetical PID control with $p = 10$ and $m = 2$.

directly to the tracking error. That is, there is no explicit computation of output reference-output measurements, rather MPC uses output measurements to update the observer of MPC. The observer is then used to predict the response of the plant across the prediction horizon.

Another important detail about the operation of MPC is that the real-time solver computes the future sequence of manipulated variables across the control horizon, but only one value is commanded to the actuators for the next time step. For the next time step, the observer values are corrected and advanced based on the sensory information and applied manipulated variables. Optimization is repeated to compute the optimal future sequence of manipulated variables across the prediction horizon.

The reader is highly encouraged to read Mathworks MPC Toolbox documentation (https://www.mathworks.com/help/mpc/getting-started-with-model-predictive-control-toolbox.html). If the link doesn't work, search for Mathworks MPC Toolbox documentation.

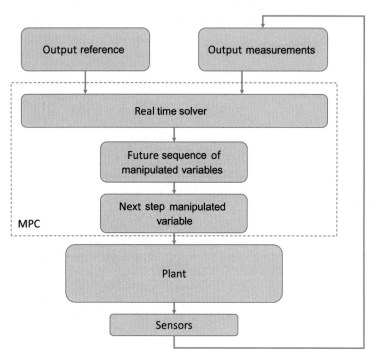

FIGURE 2.10   Simplified Block diagram of MPC.

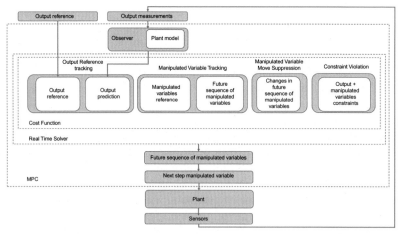

FIGURE 2.11   Detailed Block diagram of MPC.

2.6 SOLVING THE REAL-TIME OPTIMIZATION PROBLEM IN MPC

One of the main challenges of implementing MPC in real-time on hardware is computational resource. In the case of process control (such as a chemical plant) where there is a dedicated mainframe to run MPC and the process dynamics are slow, throughput is not an issue. Solving the optimization problem can take somewhere from 10 to 1000 times more than a PID or a lookup-based controller. The computation time for MPC depends on the complexity of the MPC number of output variables, manipulated variables, order of the plant, prediction horizon, control horizon, etc. Furthermore, throughput is not constant when solving for MPC. The number of iterations to find a solution varies. When embedded in a microcontroller, the worst case scenario (maximum number of iterations) needs to be known beforehand.

To mitigate throughput issues, the optimization problem can be run offline and the solution can be stored in the form of tables. What is embedded in the microcontroller are the tables. To execute MPC, the states of the system are estimated, and the commanded value for the control action is looked up from the tables. This approach is called explicit MPC, whereas running the optimization problem in the microcontroller is called implicit MPC. Whereas implicit MPC requires a significant amount of computation, but the size of the code is small, explicit MPC requires fewer computations and occupies a much bigger memory size in the hardware due to the large lookup tables. The paper by Bemporad et al., is a great starting point to get a deeper knowledge of explicit and implicit MPC [1].

2.7 MATHWORKS AND MPC

Mathworks is a leading mathematical software computing corporation that has helped the academic and industrial community advance science and technology. Mathworks embeds mature research into their toolboxes. The Model Predictive Control Toolbox is another example where Mathworks has worked closely with researchers in the field of MPC to deliver high-quality software package and documentation. The toolbox continues to improve with each release and R2017a is no different. In the toolbox, the user can design implicit MPC, explicit MPC, adaptive MPC, as well as gain-scheduled (or multiple) MPC. The focus of this book is on implicit MPC since it is a fundamental part of the rest of the techniques in the toolbox. Furthermore, implicit MPC can be used as the control solution for many industrial and academic problems. Additionally, the book will cover multiple MPC which allows the use of multiple linear MPC designs to control a nonlinear plant over a wide range of operation. Explicit MPC and adaptive MPC are not covered in this book. The reader is encouraged to read the documentation about all the techniques to gain more insight about the trade-offs between them.

REFERENCE

[1] A. Bemporad, M. Morari, V. Dua, E.N. Pistikopoulos, The explicit linear quadratic regulator for constrained systems, Automatica 38 (1) (2002) 3–20. ISSN 0005-1098.

FURTHER READING

M. Kvasnica, "Implicit vs explicit MPC—Similarities, differences, and a path owards a unified method," 2016 European Control Conference (ECC), Aalborg, 2016, pp. 603–603. https://doi.org/10.1109/ECC.2016.7810353

Chapter 3

MPC Design of a Double-Mass Spring System

3.1 INTRODUCTION

The main theme of this book is to streamline the process of designing Model Predictive Control (MPC). We used the same steps across all the applications to get to the design of the controller. One might mistakenly think that we found a way to automate the control design process and tuning for any problem using MPC. The process that is reiterated in this book gets us closer to automation of control design, tuning and deployment. The tools and MATLAB[1] functions are reused in the upcoming chapters. There are unique elements in designing a controller for different applications that hinder the automation of control design and tuning, such as:

- Time constant of the plant (milliseconds vs hours, or even days), sensors and actuators.
- Noise factors and unmeasured disturbances—these could throw off system identification processes.
- Reference generation—deciding what is the optimal trajectory is not always straightforward.
- Tuning tradeoffs for the controller.
- Nonlinearity of the plant.
- The processor that executes the controller.

In this chapter, we introduce the framework of designing and deploying MPC (Section 3.2). In the heart of this process, is the system identification which is discussed in Section 3.3. To practice system identification and MPC design processes, we introduce the popular double-mass spring plant model (Section 3.4). In Section 3.5, we perform system identification on the introduced plant. Using the developed model, we walk the reader through a step-by-step process to design a MPC controller (Section 3.6). In Section 3.7, we combine the controller with the double-mass spring model to perform a closed-loop simulation to assess the performance of the controller. An

---

1. MATLAB is a registered trademark of The MathWorks, Inc.

Practical Design and Application of Model Predictive Control.
DOI: https://doi.org/10.1016/B978-0-12-813918-9.00003-4

application problem concludes the chapter which provides a practical example of the theory.

The electronic version of all the M-script files as well as the Simulink models can be found in the *Chapter_3* folder which can be downloaded from MathWorks website. Follow the below link (file exchange) and search for the ISBN or title of the book.

https://www.mathworks.com/matlabcentral/fileexchange/

You can download the same codes, connect with authors and have access to additional free resources by following the dedicated website for the book

www.practicalmpc.com

MATLAB R2017a was used to develop the examples in this chapter. MATLAB, Simulink, MPC Toolbox, and System Identification Toolbox are required for this chapter. It is advised that the reader uses same version of MATLAB or a later version. If readers need an earlier version of the codes, please contact the authors.

3.2 MODEL-BASED DESIGN FRAMEWORK

Using the same process to design a controller for any plant model is not a trivial task. Controller design/tuning/deployment is a combination of science and art. Fig. 3.1 highlights the three main steps that are followed in the design of a MPC controller:

1. System identification
2. Controller design
3. Controller deployment

System identification is the process that generates an approximation of the plant concerned. In this book, the model is linear.

Controller design is the process of structuring and tuning a controller that meets customer requirements (see Chapter 5: Single MPC Design for a Ship for details). In this book, linear MPC is the controller that is being designed and tuned.

Controller deployment is the process of embedding the control design into real-time hardware. This part is covered in the Chapter 9 of the book.

3.3 SYSTEM IDENTIFICATION PROCESS

Model-based controllers (such as, MPC, state feedback control, linear quadratic regulators, etc.) require a model that captures the main dynamics of the plant in the region of operation where the plant is to be controlled.

FIGURE 3.1 Controller design and deployment framework.

Frequently asked questions include: How accurate should the model be? Is a linear approximation good enough in the region of operation? This section will tackle these questions and outline the process for obtaining a good model.

The inputs in the diagram Fig. 3.2 are the independent variables and can be classified into measured disturbance, unmeasured disturbance, and manipulated variables (MVs). The measured disturbances are inputs that cannot be adjusted by the controller. For example, wind speed acting on a helicopter can be measured, but cannot be manipulated or changed by the controller. Unmeasured disturbances are inputs that the controller is unaware of, but they affect the response of the system. For example, wind resistance affects the speed of a car, but the cruise control logic of the speed of a car doesn't have wind resistance as a measurement. MVs are the levers that the controller changes to affect the response of the plant. In the case of a ship, the controller manipulates the rudder angle to change the heading angle of the ship.

Outputs are the dependent variables (responses) of the plant. Outputs can be classified into measured and unmeasured outputs. The measured outputs are used directly or indirectly (through estimation of other outputs) by the controller to assess how far the actual response of the plant is from the desired response. For example, in the case of cruise control for a car, the driver wants to maintain a constant speed—i.e., desired speed. The vehicle speed sensor provides the controller with a direct feedback to the current speed. In this case, vehicle speed is a measured output. On the other hand, the motion of the car in the vertical direction is an unmeasured output—in general no sensory information is available for the vertical vibration of a car.

There are mainly two ways to obtain linear approximation of a plant:

- Input/output data-driven linear approximation.
- Simplification of nonlinear equations of motions.

In the context of this book, the input/output data-driven approximation of a plant will be used to perform system identification. The input/output data can be either time series data or frequency-domain data. Here, the former will be used. Fig. 3.3 shows the available measurements (dashed arrows) that

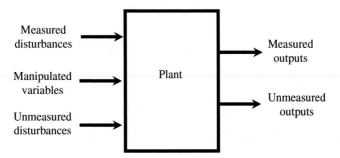

FIGURE 3.2 Detailed diagram of a plant with its inputs and outputs.

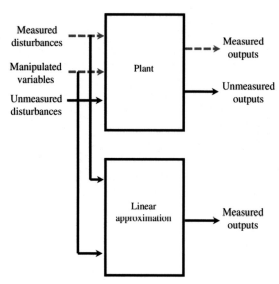

FIGURE 3.3 Data-driven linear approximation of a plant.

can be used to identify the plant. The measured disturbances along with the MVs will be used as inputs for the linear approximation. The outputs of the linear approximation will be compared with the measured outputs of the plant to judge the quality of the approximation.

The general guidelines to obtain good input/output data that are required to capture the main dynamics of a plant model are:

- The MVs should be adjusted enough to generate a reasonable response of the plant (measured output).
- The noise-to-signal ratio in both the input and output should be small.
- The unmeasured disturbance should not impact the measured outputs more than the MVs.

Based on the authors' industrial and academic experience, the value of the linear approximation of a plant is highly underestimated and often assumed to be a trivial exercise that every control engineer and student can, and should, know how to do. In practice, obtaining a useful linear approximation is a nontrivial especially when dealing with black-box plants with an unknown structure, nonlinear plants, or noisy measurements, in addition to unmeasured inputs and disturbance.

Having a highly capable numerical computing environment such as MATLAB helps to reduce the difficulties of the above task by providing specialized toolboxes (such as the System Identification Toolbox). But the usage of such software doesn't eliminate the need for a solid understanding of the plant, its measured and unmeasured inputs and outputs, as well as the nonlinearity of the plant concerned.

These are questions that the authors are often asked by control designers:

1. What kind of data do I need to collect for my system identification and linearization of the plant?
2. I can't seem to get a good linear model approximation for my plant. What am I doing wrong?
3. What should the order of the linear model approximation be?
4. The linear model seems to agree well with the plant in some regions while in others it doesn't. Why?
5. While conducting an experiment to collect data for system identification, I moved all my MVs, but it seems that the plant response to some of them was negligible or even nonexistent. What did I do wrong?

With regard to the second question, there may be several causes that could lead to an unacceptable linear approximation. It may be that the order of the linear model is not proper, in which case the reader needs to redo the identification with a different order—it might take few iterations of trials, but the general guideline is to start with a low order and move upwards. In some cases, it might be that the MVs haven't been changed enough to generate a reasonable response in the plant and thus the system identification has numerical problems in getting a good linear approximation. Another cause for the problem of getting a bad linear approximation might be that the region of operation where the user is performing the approximation might be highly nonlinear. This might cause issues in the linearization especially if the plant is experiencing a sign change in the gain of plant. Fig. 3.4 shows the steady state output response of a plant as a function to the actuator response. The plant is nonlinear. Moreover, as indicated in the two regions

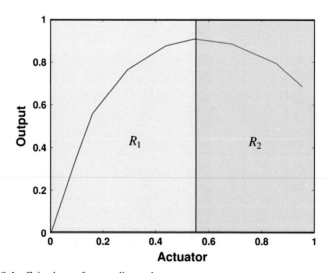

FIGURE 3.4 Gain change for a nonlinear plant.

on the plot, the gain is positive in region 1 whereas it is negative in region 2. Approximating the plant across both regions R_1 and R_2 using one linear model will lead to an inaccurate linear model.

As for the order of the linear model in question three, the reader should start with a low-order model and increase the order if an adequate fit or accuracy is not reached.

Question four is particularly tricky. The reader must realize that linear models are approximating models and are adequate in a specific region of operation especially if the plant is highly nonlinear. Understanding a nonlinear plant before attempting to linearize it, is a crucial step in achieving a reasonable approximation. In many cases, one linear model is not sufficient to approximate the plant across all the regions of operation. Fig. 3.5 shows an example of where five linear plants are needed to approximate a nonlinear plant.

As for question five, one of the causes for a nonlinear plant's negligible response to one of the MVs is that the nonlinear plant might not be sensitive to that specific MV in that specific region of operation. Fig. 3.6 shows the steady state response of a nonlinear plant as a function of the actuator. The figure shows two regions. The one on the right shows that the plant's response is not sensitive to the actuator.

Input and output time series data for a plant contain valuable information about the dynamics of the system. It is not always feasible to obtain a linear approximation of a plant from any set of input and output time series data. Properly planning an experiment where the plant is reasonably fired up would alleviate system identification problems. The process of obtaining a linear approximation of a plant is outlined in Fig. 3.7:

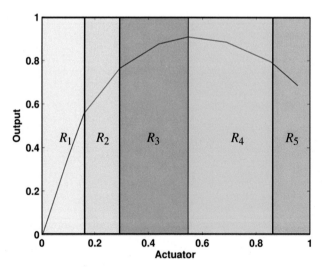

FIGURE 3.5 Nonlinear plant with five approximating linear plants.

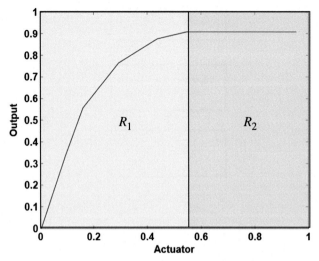

FIGURE 3.6   Nonlinear plant with a flat response to an actuator in region R_2.

1. Design of experiment: In preparation for this step, measured disturbances, unmeasured disturbances, MVs, as well as measured and unmeasured outputs should be identified. The range of the MVs should be defined. Also, if there is any previously recorded data for the plant, the reader can (coarsely) identify the resolution of the response of the plant to changes in the MVs. For example, the change in speed of a car can be attributed to change in the accelerator pedal angle (e.g., 5 degrees change in the accelerator pedal would cause 10 km/h change in the speed of the car). In this step, the reader needs to know the area of operation of the MVs where the plant will be linearized. The MVs will be scheduled to move, as a function of time, in the area of operation. One way of scheduling the MVs is to step them one MV at a time and allowing enough time for the plant to reach a steady state, then step to the next MV, and so on.
2. Implement experiment and collect data: In this step, the design of the experiment is implemented and the relevant parameters are logged—preferably at ten times faster than the time constant of the plant.
3. Using the collected data, the user searches for the best-fitting linear model.
4. The user utilizes the fitting error to judge if the fitted linear model is good enough. Moreover, the user leverages his/her expert knowledge about the plant concerned to judge if the obtained linear model is acceptable. For example, for a stable plant, the user can check whether the linear model is stable. If the user is not satisfied with the fit, he/she might need to go back to the design of the experiment, especially if the plant response was poor.
5. Using a new set of input/output plant data that wasn't used for the fitting process, the user tests the linear model. This data is referred to as validation data.
6. If the validation error is unacceptable, the user must go back to step one.

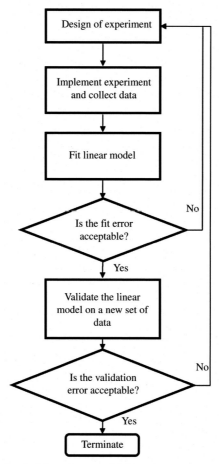

FIGURE 3.7 Process for obtaining a linear model for a plant.

One tip that could help the reader get a good linear model is to try to start with a first order system. If the first order system doesn't yield a good fit, then the reader can increase the order of the system until a good fit is reached. When searching for a linear fit for the plant, follow Einstein's advice: "Everything should be made as simple as possible, but no simpler."

3.4 DOUBLE-MASS SPRING SYSTEM

The plant model that will be used to exercise system identification and MPC controller design is the double-mass spring system shown in Fig. 3.8. The two mass points m_1 and m_2 are connected by three springs with spring constants k_1, k_2. In this model, it is assumed that there is no friction between the blocks and the floor. There are two input forces u_1 and u_2 applied on the blocks. These forces must overcome the spring forces to displace the masses.

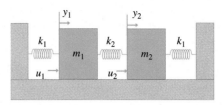

FIGURE 3.8   Double-mass spring system.

The outputs y_1 and y_2 are the displacements of the blocks due to the input forces u_1 and u_2.

Applying Newton's second law, the state equations of the system can be obtained.

$$u_1(t) - k_1 y_1(t) - k_2(y_1(t) - y_2(t)) = m_1 \ddot{y}_1(t)$$
$$\ddot{y}_1(t) = \frac{1}{m_1} u_1(t) - \frac{(k_1 + k_2)}{m_1} y_1(t) + \frac{k_2}{m_1} y_2(t) \tag{3.1}$$

Similarly, for the second equation of motion:

$$u_2(t) + k_2 y_1(t) - (k_1 + k_2) y_2(t) = m_2 \ddot{y}_2(t)$$
$$\ddot{y}_2(t) = \frac{1}{m_2} u_2(t) + \frac{k_2}{m_2} y_1(t) - \frac{(k_1 + k_2)}{m_2} y_2(t) \tag{3.2}$$

Let us define the states of the system to be:

$$x_1(t) := y_1(t)$$
$$x_2(t) := \dot{y}_1(t)$$
$$x_3(t) := y_2(t)$$
$$x_4(t) := \dot{y}_2(t)$$

Thus, the state space equation can be written as:

$$
\begin{bmatrix} \dot{x}_1(t) \\ \dot{x}_2(t) \\ \dot{x}_3(t) \\ \dot{x}_4(t) \end{bmatrix}
=
\begin{bmatrix}
0 & 1 & 0 & 0 \\
-\dfrac{(k_1 + k_2)}{m_1} & 0 & \dfrac{k_2}{m_1} & 0 \\
0 & 0 & 0 & 1 \\
\dfrac{k_2}{m_2} & 0 & -\dfrac{(k_1 + k_2)}{m_2} & 0
\end{bmatrix}
\begin{bmatrix} x_1(t) \\ x_2(t) \\ x_3(t) \\ x_4(t) \end{bmatrix}
+
\begin{bmatrix}
0 & 0 \\
\dfrac{1}{m_1} & 0 \\
0 & 0 \\
0 & \dfrac{1}{m_2}
\end{bmatrix}
\begin{bmatrix} u_1(t) \\ u_2(t) \end{bmatrix}
$$

And the output:

$$y(t) := \begin{bmatrix} y_1(t) \\ y_2(t) \end{bmatrix} = \begin{bmatrix} 1 & 0 & 0 & 0 \\ 0 & 0 & 1 & 0 \end{bmatrix} x(t) \tag{3.3}$$

Below are the physical properties of the system:

Mass $m_1 = 1\,\text{kg}$
Mass $m_2 = 1.2\,\text{kg}$
Spring Constant $k_1 = 1\,\text{N/m}$
Spring Constant $k_2 = 1.3\,\text{N/m}$

MATLAB code that shows the implementation of the state space model in Eq. (3.3) is as follows:
Chapter_3_Section_4_Script_1.m

```
%Book Title: Practical Design and Application of MPC
%Chapter: 3
%Section: 4
%Authors: Nassim Khaled and BibinPattel
%Last Modified: 10/15/2017
%%
clc
clear all
bdcloseall
m1 = 1;       %kg
m2 = 1.2;     %kg
k1 = 1;       %N/m
k2 = 1.3;     %N/m
%% State Space Matrices
A = [0              1    0              0;
     -(k1+k2)/m1    0    k2/m1          0;
     0              0    0              1;
     k2/m2          0    -(k1+k2)/m2    0];

 B = [0       0;
      1/m1    0;
      0       0;
      0       1/m2];

C = [1  0   0   0;
     0  0   1   0];

 D = [0       0;
      0       0];

%% Create the Continuous time State Space Model.
Continuous_Plant_Model=ss(A,B,C,D);
% Set the Inputs and Output names of the model with units
Continuous_Plant_Model.InputName = {'u1','u2'};
Continuous_Plant_Model.InputUnit={'N','N'};
Continuous_Plant_Model.OutputName = {'y1','y2'};
Continuous_Plant_Model.OutputUnit={'m','m'};

fprintf('Continuous Time Plant Model \n');
Continuous_Plant_Model
%% Create the Discrete Time Plant Model. c2d function converts the
Continuous time model to discrete time. User needs to specify
% the sampling time of the system.
%% Save the Model
save Continuous_Plant_Model.matContinuous_Plant_Model
```

Fig. 3.9 shows the MATLAB command window outputs after executing *Chapter_3_Section_4_Script_1.m*

Continues time plant model

Plant_model =

a =

	x1	x2	x3	x4
x1	0	1	0	0
x2	−2.3	0	1.3	0
x3	0	0	0	1
x4	1.083	0	−1.917	0

b =

	u1	u2
x1	0	0
x2	1	0
x3	0	0
x4	0	0.8333

c =

	x1	x2	x3	x4
y1	1	0	0	0
y2	0	0	1	0

d =

	u1	u2
y1	0	0
y2	0	0

FIGURE 3.9 State space model of the mass spring system.

Now we will generate the Simulink model using the state space equations. Follow the below steps.

1. Create a new Simulink model and save it as:
   Mechanical_Mass_Spring_Continuous_Time.slx as shown in Fig. 3.10
2. Open the Simulink library and browse to Continuous blocks and add a state space block into the newly created model as shown in Fig. 3.11
3. Name the newly added state space block as *Continuous Mass Spring Model*, as shown in Fig. 3.12.
4. Double click on the state space block added to the model and enter the A, B, C, D parameters as *Continuous_Plant_Model.a,*
   Continuous_Plant_Model.b, Continuous_Plant_Model.c,
   Continuous_Plant_Model.d as shown in Fig. 3.13.
5. Since this is a 2 input 2 output system, add the step blocks for the inputs and scopes for the outputs as shown in Fig. 3.14. Also add two scopes for logging the step input data. Apply a step change from 0 to 1 to the force u_1 at time 50 seconds. Similarly apply a step change from 0 to 1 to force u_2 at time 150 seconds.
6. Configure the model's simulation parameters using the *Simulation >> Model Configuration Parameters* menu. Since this is a continuous time

FIGURE 3.10   Creating new model.

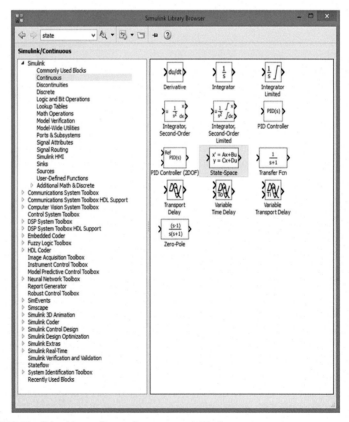

FIGURE 3.11   Selecting continuous time state space block.

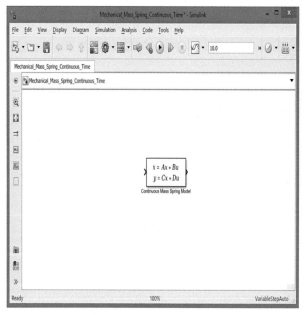

FIGURE 3.12   State space block added to the new model.

FIGURE 3.13   Configuring the state space block parameters.

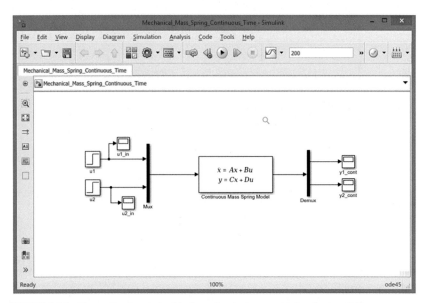

FIGURE 3.14 Connecting input step blocks and output scopes to the plant model.

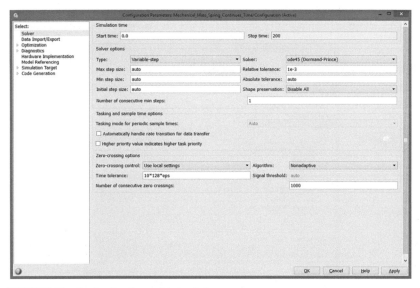

FIGURE 3.15 Configuring the model simulation parameters.

model, we will use the variable-step solver and let the Simulink engine figure out the time step to run the model. Change the simulation start time to be 0 and end time to be 200 seconds and click Apply. Shown in Fig. 3.15.

7. Simulate the model and observe the open loop response of the system.

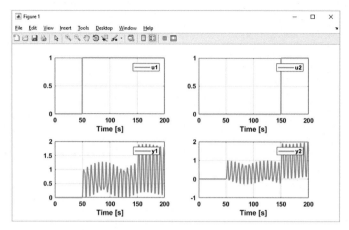

FIGURE 3.16 Open loop simulation.

Use the below code snippet to plot the data. Fig. 3.16 shows the open loop response of the model for each input against the output.

To regenerate the results in Fig. 3.16, the reader can run *Chapter_3_Section_4_Script_2.m* script which can be found in the *Chapter_3/Section_4* material.

Chapter_3_Section_4_Script_2.m

```
%Book Title: Practical Design and Application of  MPC
%Chapter: 3
%Section: 4
%Authors: Nassim Khaled and BibinPattel
%Last Modified: 10/15/2017
%%
clc
clear all
bdcloseall
close all
% Load the Continuous Time Model of Mass Damper System
load Continuous_Plant_Model.mat
% Open the simulink model
open_system('Mechanical_Mass_Spring_Continuous_Time');
% Simulate the Model
sim('Mechanical_Mass_Spring_Continuous_Time');
%% Plot the results
subplot(221)
plot(u1_in(:,1),u1_in(:,2),'linewidth',2);
hold all
grid on
xlabel('Time [s]');
legend('u1');

subplot(222)
plot(u2_in(:,1),u2_in(:,2),'linewidth',2);
hold all
grid on
xlabel('Time [s]')
legend('u2');
```

```
subplot(223)
plot(y1_cont(:,1),y1_cont(:,2),'linewidth',2);
hold all
grid on
xlabel('Time [s]');
legend('y1');

subplot(224)
plot(y2_cont(:,1),y2_cont(:,2),'linewidth',2);
hold all
grid on
xlabel('Time [s]');
legend('y2');
set(gcf,'color',[1 1 1]);
```

3.5 SYSTEM IDENTIFICATION FOR A DOUBLE-MASS SPRING PLANT

In practical applications, equations that accurately describe the dynamics of the plant are not available. Thus, we revert to experiments where we fire up the plant model with given inputs, record the outputs and run the system identification process outlined in Fig. 3.7. For the double-mass spring plant, we will use the data collected in the previous section (Fig. 3.16).

Chapter_3_Section_5_Script.m simulates the Simulink plant model generated in the previous section and performs the system identification using the *MATLAB System Identification* toolbox commands. The function *iddata* is used to prepare the data in the proper format for system identification. The system identification function *n4sid* is used to generate an initial linear model. The resultant linearized model is further refined using the *pem* function. To compare the results of the identified model to the original measurement data, the function *compare* is used. The output data from *compare* is used to generate a comparison plot between the actual data and the linear model. Results of the system identification are saved under the name *Chapeter_3_Section_5_Sys_ID_Model.mat*.

From the state space model derived earlier, it was known that this is a system with order 4. This was used as an input to *n4sid*. Identifying the order of the system is not always straightforward. Start from an order of 1 and then perform the system identification. Increment the order to get the best fit.

Chapter_3_Section_5_Script.m can be found in *Chapter_3/Section_5* folder.

Chapter_3_Section_5_Script.m

```matlab
%Book Title: Practical Design and Application of MPC
%Chapter: 3
%Section: 5
%Authors: Nassim Khaled and BibinPattel
%Last Modified: 10/15/2017
%% Load the discrete time mass damper system and run the system ID simulation
clear all
bdcloseall
close all
% Load the Continuous Time Model of Mass Damper System
load Continuous_Plant_Model.mat
% Open the model and simulate
open_system('Mechanical_Mass_Spring_Continuous_Time');
sim('Mechanical_Mass_Spring_Continuous_Time');
%% Grouping simulation data and neglecting first few seconds(Nstart*stp_sz)
stp_sz = 0.1;
Nstart = 5; %Clip the data starting from Nstart as the first few seconds of the
simulation should be discarded
Measured_Outputs=[y1_cont(Nstart:end,2) y2_cont(Nstart:end,2)]; %Measured outputs
Manipipulated_Variables=[u1_in(Nstart:end,2) u2_in(Nstart:end,2)] ; %Manipulated
variables

%% Obtaining initial conditions at step time=Nstart
Measured_Outputs_Nstart=Measured_Outputs(Nstart,:); %Capturing the measured outputs at
step time= Nstart
Manipipulated_Variables_Nstart=Manipipulated_Variables(Nstart,:) ; %Capturing the
manipulated variables at step time= Nstart

%% Forcing response to start from zero initial conditions
Measured_Outputs_zero_initial_conditions=Measured_Outputs-
repmat(Measured_Outputs_Nstart,length(Measured_Outputs),1); %Subtracting initial
conditions for measured outputs to obtain zero response at step time= Nstart
Manipipulated_Variables_zero_initial_conditions=Manipipulated_Variables-
repmat(Manipipulated_Variables_Nstart,length(Manipipulated_Variables),1); %Subtracting
initial conditions for manipulated variables to obtain zero actuation at step time=
Nstart

%% Prepare date for system identification
data=iddata(Measured_Outputs_zero_initial_conditions,Manipipulated_Variables_zero_initia
l_conditions,stp_sz); %data is packaged for system identification using idda-ta

%% Generate a preliminary 4th order system that fits the data
system_order = 4;
sys1=n4sid(data,system_order,'Form','canonical','DisturbanceModel','None','InputDelay',
[0 0]','InitialState','zero'); %n4sid generates a preliminary system in the canonical
form
%with zero disturbance, zero delay and zero initial conditions

%% Generate a more refined system
sys2 = pem(data,sys1,'InitialState','zero') %pemgenerates a more refined system that
fits the data better

%% Define the options for comparing the various identified systems
opt = compareOptions('InitialCondition','Z');
[Y,fit,x0]=compare(data,sys2);
Y_1=Y.OutputData(:,1);
Y_2=Y.OutputData(:,2);
%% Plot the Measured and Modeled Outputs
figure
subplot(211);
plot(0:stp_sz:stp_sz*(length(Y_1)-
1),Measured_Outputs_zero_initial_conditions(:,1),'linewidth',2)
grid on
hold on
plot(0:stp_sz:stp_sz*(length(Y_1)-1),Y_1,'--r','linewidth',2)
legend('y1 Measured','y1 System Identification')
title(['System ID Order ' num2str(system_order)]);
xlabel('Time [s]')
subplot(212);
plot(0:stp_sz:stp_sz*(length(Y_2)-
1),Measured_Outputs_zero_initial_conditions(:,2),'linewidth',2)
```

```
grid on
hold on
plot(0:stp_sz:stp_sz*(length(Y_2)-1),Y_2,'--r','linewidth',2)
legend('y2 Measured','y2 System Identification')
set(gcf,'color',[1 1 1 ]);
xlabel('Time [s]')

%% Save the system into a mat file
Chapeter_3_Section_5_Sys_ID_Model = sys2;
save Chapeter_3_Section_5_Sys_ID_ModelChapeter_3_Section_5_Sys_ID_Model
```

Fig. 3.17 shows results of system identification. In this case, the identi-fied model perfectly agrees with the plant model (100% fit). This is a special case as the plant model is a linear one. In practical applications, the plant will be nonlinear with unmeasured disturbance. A fit greater than 75% can be challenging for some plants.

3.6 MPC CONTROL DESIGN

In this section, we will design a MPC controller to force the position of the two masses to follow a prescribed trajectory. The process of designing MPC in MATLAB is well-packaged in the *MPC Designer App*. Follow the steps below to design the controller:

1. To open the *MPC Designer App* type *mpcDesigner* in the MATLAB command window (Fig. 3.18).
2. The GUI has three submenu lists under *MPC Design Task* Menu list. The *Plant Models* menu is where you can import plant models, the *Controllers* menu allows you to select the imported plant model/models to develop the MPC controller, specify the constraints, and allows

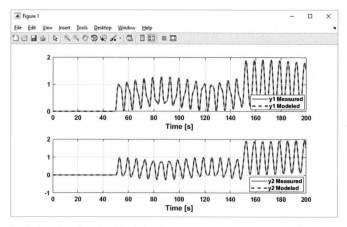

FIGURE 3.17 Results of system identification.

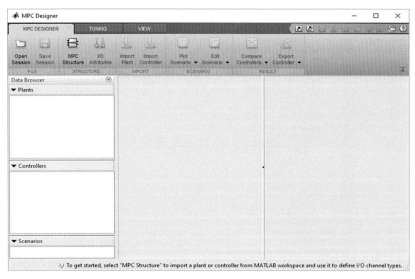

FIGURE 3.18   MPC *Designer App* GUI.

weight tuning, etc. And the *Scenarios* menu is where you can test the designed controller with the plant model to make sure the controller is working as expected. If not, you can iterate by going back to the *Controllers* menu, redesign it and test again.

3. Let's start by importing the plant model first. Click on the button *MPC Structure* on the GUI. This will pop up another GUI as shown in Fig. 3.19. You can see that when this GUI is opened it will search in the workspace for any available state space models and display it for the user. After that, click on the row to select the plant model you want to import to the MPC Design GUI. Select *Chapter_3_Section_8_Sys_ID_Model*. It will show the properties of the selected plant. Click the *Define and Import* button and the designer will create a default MPC controller structure with the imported plant, controller, and scenario as shown in Fig. 3.20.

4. The main GUI will now be populated with the plant you imported as shown in Fig. 3.20. It will show the inputs of the system u_1 and u_2 as MVs and outputs y_1 and y_2 as measured variables. Under the *Controllers* menu, there is a default controller called *mpc1* added automatically by the *Define MPC Structure* GUI. The user can change the name of this controller to any name you want—click on the name of the controller three times to display the name in an edit box to edit the name. Also, it will add Scenario1 as shown—this can be renamed as well. In this example, we renamed the controller to be *Mass_Spring_MPC* and the Scenario1 to be *Mass_Spring_Test*. When the MPC structure is created,

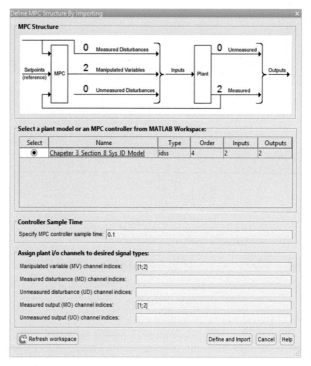

FIGURE 3.19   Plant model importer GUI.

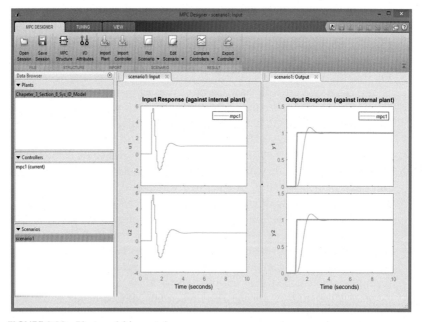

FIGURE 3.20   Plant model imported.

it will try to simulate the default controller with the default scenario and plot the results as shown in Fig. 3.20.

5. Now click on the *Tuning* tab at the top of the *MPC Designer GUI*. The controller design window will be displayed at the top of this window. Here you can select the controller and plant from the drop-down menu. The drop-down menu to select a plant model can be used if the user has multiple plant models to select from. You will learn more about this in chapter 6 when we deal with multi-mode MPC controller design. Select the *Chapter_3_Section_8_Sys_ID_Model* that was imported earlier. After selecting the plant, you need to specify the sample time of the controller. The GUI automatically takes the sample time of the imported plant as the *Sample Time* (in seconds) as shown. This can be changed to a different value if you want to run the controller at a different rate than that of the plant model—additional rate transition blocks might be needed to handle it when you run the simulation in Simulink. In this case, you will use the same control interval as that of the plant. Next are the Prediction and Control horizon intervals. By default, the GUI will use a prediction horizon of 10 intervals which is 1 second and a control horizon of 2 intervals which is 0.2 seconds according to the control interval (controller sample time)—we have selected as 0.1 seconds. We will keep the default prediction and control horizon intervals as it is and observe the performance of the controller and retune it if it does not meet the desired performance. See Fig. 3.21.

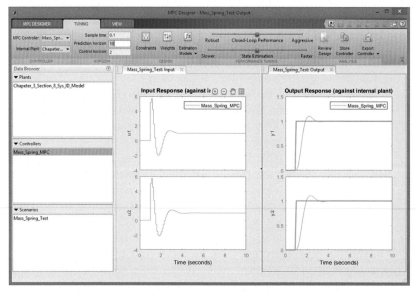

FIGURE 3.21   Selecting the model and controller horizons.

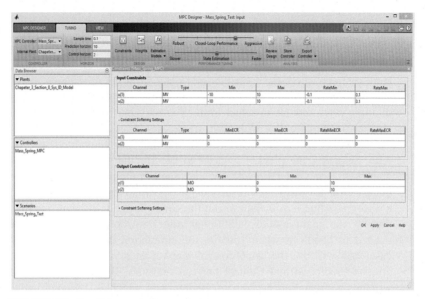

FIGURE 3.22 Specifying controller constraints.

6. Now let us move to the next option, *Constraints*—shown in Fig. 3.22. Click on the *Constraints* button on the ribbon to specify the input and output constraints the MPC controller design needs to meet. You can specify the constraints for both input and output variables separately here. For the input MVs, the Min and Max allowed values, and the rate of increment and decrement for the control action can be constrained by entering values into the cells provided. In this example, we are constraining the input forces to range from -10 to $10\,\text{N}$ fall and the rise and fall rate of the control action to be 0.1 N. Similarly, the output displacement is constrained between 0 and 10 m.

7. Now let us move to the next menu option, *Weight Tuning*—shown in Fig. 3.23. Click on the *Weights* button on the ribbon to specify the input and output weights. This section allows us to adjust the controller performance by specifying weights on the input MVs, the outputs, or both. The *Input Weights* table lists all the MVs of the controller. For each MV, we can specify the *Weight* and *Rate Weight* options. The *Weight* option penalizes the change in the MV from its nominal value specified (see Fig. 3.23). We can keep this cell either 0 or blank so as not to penalize the control action. This means that the MV can move anywhere between its lower and upper bounds provided in the *Constraints* option. Or, you can give a positive value for the *Weight* to penalize the MV to remain closer to its nominal value. In this case, we kept the weight as zero for both the MVs. The *Rate Weight* option penalizes the change in value of a MV. You can keep this cell either 0 or blank not to penalize

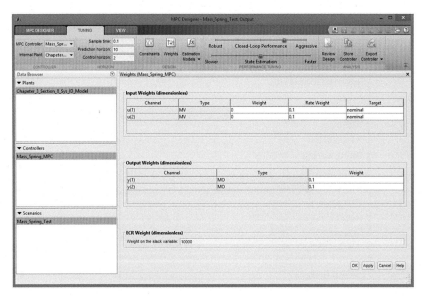

FIGURE 3.23 Specifying controller weight tuning.

the rate change, the MV can move up or down at any rate within the bounds provided in the *Constraints* option. A positive value for the *Rate Wight* applies a penalty to reduce the rate of change of MV. Keep the default value as 0.1 for the *Rate Weight* in this case. Similarly, the *Input Weights* table lists all the outputs of the controller. For each output, we can specify the *Weight* option. The *Weight* option penalizes the deviation of the output from its set point. You can keep this cell either zero or blank so as not to apply penalty when the output deviates from the set point. Applying a positive value adds a penalty to keep the output near the set point. Keep this output weight at 0.1.

8. In this example of the controller design, we are not using any specific estimation models for the plant output disturbance or sensor measurement noise. So, do not make any changes to the default setting of *Estimation Models* option.

9. The next option in the controller design is to adjust the importance between the weights of the manipulated or output variables to the MV rate weights. The *Closed Loop Performance* slider in the GUI adjusts the weights on all the input and output variables. Moving the slider to the left increases the MV rate penalties relative to set point penalties, often increasing controller robustness, but with a slower response to disturbances. In contrast, moving the slider to the right makes the controller more aggressive, but less tolerant toward modeling errors. Keep the slider at its default value of 0.5. The slider for the *State Estimation* is left to use the default value as well, since we have not made any

FIGURE 3.24   Hard MV constraints warning.

changes to the default setting of the plant output and measurement model in Step 8.

10. After the controller design, we can perform a design review to examine the MPC controller for design errors and stability problems at runtime just by clicking the option *Review Design* in the *MPC Designer* GUI. This will open a HTML report as shown in Fig. 3.24. You will see that all the controller performance tests passed except the *Hard Manipulated Variable Constraints*. The user can click on the link to get a better explanation on the warning. In this case, the warning is because you put hard constraints on both the MVs and their rates at the same time. To handle this problem, you can "soften" the constraints. The quadratic solver of MPC, if it computes an unfeasible solution, will sacrifice the softened constraints to find a feasible solution. You can do this by going back to the *Constraints* option in the *MPC Designer* GUI. Expand the *Constraint Softening Settings* and specify a non-zero value (1 in this example) for the *RateMinECR* and *RateMaxECR* for both the MVs as shown in Fig. 3.25. Repeat the *Review Design* option to make sure that there are no warnings or errors in the MPC design (Fig. 3.26).

11. At this point, you are ready with an initial design of the controller and can move on to test the controller performance with a simple reference input for the controller to track. Go to the *MPC Designer* tab and click on *Edit Scenario* and select the scenario *Mass_Spring_Test*. A window will be displayed as shown in Fig. 3.27. In the scenario window, select

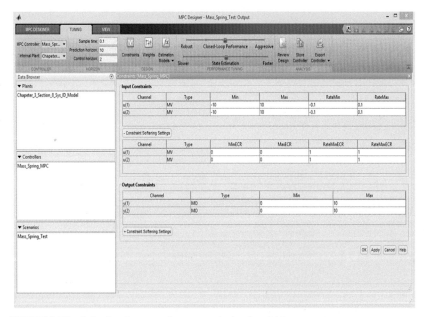

FIGURE 3.25   Softening the constraints on manipulated variable rates.

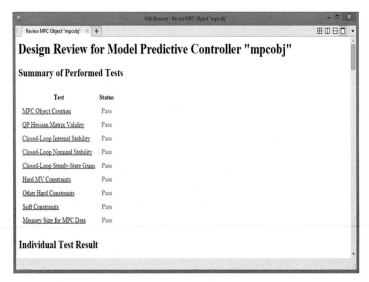

FIGURE 3.26   MPC design review all performance tests passed.

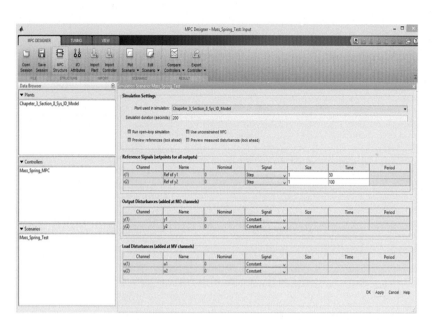

FIGURE 3.27   Editing the test scenario to validate MPC controller.

the plant model to be *Chapter_3_Section_8_Sys_ID_Model*, and the simulation duration to be 200 seconds. Under the reference signals section in the *Signal* column, select *Step* input for both output references y_1 and y_2. With a default nominal value of 0 for the step input, change the step size to be 1 for the references, but change the time at which the step happens for y_1 and y_2. For y_1 let it be at 50 seconds and for y_2 step at 100 seconds. Do not add any output or load disturbances in this example and click on the *Apply* and the *Ok* buttons. The MPC Designer tool will now simulate the selected plant model with the MPC controller designed in the above steps along with the reference tracking outputs defined in the scenario window and plot the MPC controller commands and the MPC controller performance tracking as shown in Fig. 3.28.

12. You can change the various settings in the *Tuning* tab such as *Constraints*, *Weights*, *Prediction* and *Control Horizon*, etc., and see the controller performance with the same scenario instantly. After satisfactory validation of the controller, you can save the controller design project into a mat file with all the settings in the *MPC Designer* GUI. This will help the users to open the controller design later and make adjustments as desired. Use the *Save Session* button to open the GUI to save the design session into a mat file. This example design is saved as *MPC_DesignTask_Chapter_3_Section_6.mat* and can be found in the *Chapter_3/Section_6* folder.

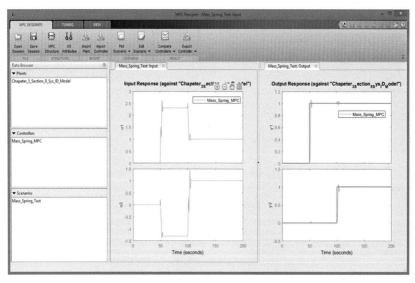

FIGURE 3.28   MPC controller performance tracking with test scenario.

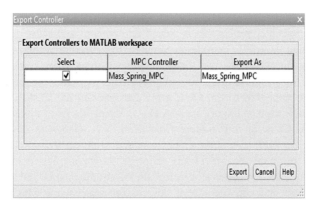

FIGURE 3.29   Exporting the MPC controller into MATLAB workspace.

13. You can also export only the MPC Controller into the MATLAB work-space using the *Export Controller* option. Fig. 3.29 shows the *Export Controller* GUI. The user will be able to change the name of the controller object while exporting, if needed, by changing the name in the *Export As* column. After exporting the controller into the MATLAB workspace, you can type in the controller object name on the MATLAB command window to see the structure and properties of the controller object as shown in Fig. 3.30.

```
Command Window
>> Mass_Spring_MPC

MPC object (created on 27-May-2017 10:57:30):
---------------------------------------------------
Sampling time:      0.1 (seconds)
Prediction Horizon: 10
Control Horizon:    2

Plant Model:
                                 --------------
        2  manipulated variable(s)  -->| 4 states  |
                                     |              |--> 2 measured output(s)
        0  measured disturbance(s)   -->| 2 inputs  |
                                     |              |--> 0 unmeasured output(s)
        0  unmeasured disturbance(s) -->| 2 outputs |
                                 --------------
Disturbance and Noise Models:
         Output disturbance model: user specified (type "getoutdist(Mass_Spring_MPC)" for details)
         Measurement noise model: user specified (type "Mass_Spring_MPC.Model.Noise" for details)

Weights:
         ManipulatedVariables: [0 0]
     ManipulatedVariablesRate: [0.0353 0.0353]
              OutputVariables: [0.2829 0.2829]
                         ECR: 10000

State Estimation:  Default Kalman Filter (type "getEstimator(Mass_Spring_MPC)" for details)

Constraints:
 -10 <= u1 <= 10, -0.1 <= u1/rate <= 0.1, 0 <= y1 <= 10
 -10 <= u2 <= 10, -0.1 <= u2/rate <= 0.1, 0 <= y2 <= 10
>> save Mass_Spring_MPC.mat Mass_Spring_MPC
fx >> |
```

FIGURE 3.30 Exported MPC controller object in MATLAB workspace.

This exported controller can be saved into a mat file and be used later for integrating the MPC controller in Simulink (discussed in Section 3.7).

14. All the above steps we have followed above can be done using MATLAB programming. Expert users tend to prefer the programming method compared to the GUI. *MPC Designer* in MATLAB 2017a provides an option to export the MATLAB script file from the MPC design session. Click on the *Export Controller* >> *Generate Script* option. Select the MPC Simulation Scenario and click on the *Generate Script* button to create the MATLAB script as shown in Figs. 3.31 and 3.32. You can save the generated script under another name and run it from MATLAB and can see that it designs the exact same MPC controller and simulates the scenario specified in the *MPC Designer* GUI.

3.7 INTEGRATING MPC WITH SIMULINK MODEL

In this section, the MPC controller designed in Section 3.6 will be integrated with Simulink. *MPC Designer* GUI in MATLAB 2017a provides an easy way to develop the Simulink model structure. It is always better to start with this sample framework of the Simulink model structure and make the necessary changes later. In order to get the basic Simulink Controller + Plant + Controller tracking reference structure go to the *Export Controller* >> *Generate Simulink Model* option in the *MPC Designer* GUI. A Simulink model with a *Step* Input

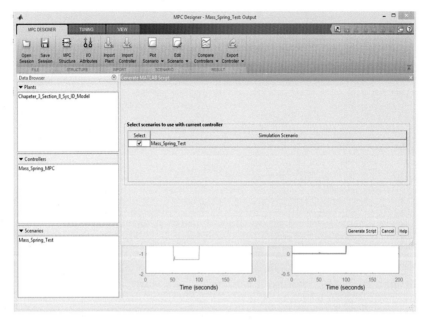

FIGURE 3.31   GUI option to generate MATLAB script to design MPC controller.

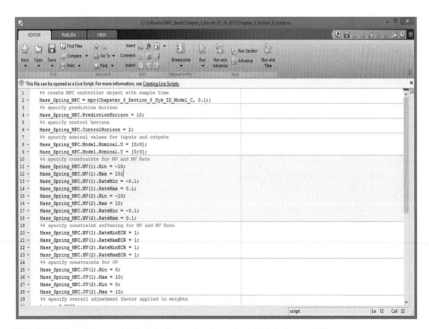

FIGURE 3.32   Generated MPC design script from the MPC designer GUI.

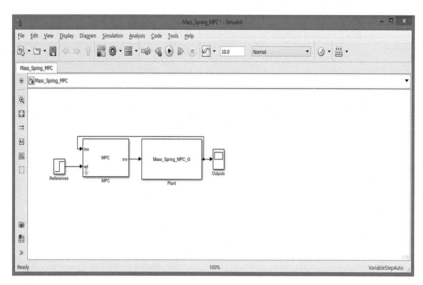

FIGURE 3.33 Generated Simulink controller framework from MPC designer GUI.

reference block, a Simulink *MPC Controller* block, and a *Linear Time Invariant* block will be added to a Simulink model (see Fig. 3.33).

We will use this exported framework model, but change the controller to track a different reference. Delete the references *Step* input block and add two *Step* Input blocks from the Simulink library. These step input blocks will be used for feeding the target reference for the displacements of the controller to track. Let us give the initial displacement target value to be 0.5 m for y_1 and a step change to 0.75 m at time 50 seconds. Do the same for y_2. Give an initial value of 0.6 m and step change to 0.3 at time 100 seconds. Connect these step input blocks to the MPC Controller *ref* input port through a Mux block. Connect the scopes for logging the reference signals, control action, and plant output as shown in Fig. 3.34. Double click on the *MPC Controller* block and use the *Mass_Spring_MPC* exported from the *MPC Designer* GUI in Section 3.6. Double click on the *Plant* block and use the *Continuous_Plant_Model* developed in Section 3.4.

One essential step when deploying a controller in a closed-loop system is to understand the execution rate of the controller and the sampling frequency of the sensors. Set the step size of the Simulink model to 0.1. This is the same step size of the MPC controller. This means that the feedback will be sampled at the same frequency as the Simulink model of 1 reading for every 0.1 s. Even though it is not needed for this exercise, it is a good practice to add rate transition blocks as a reminder that the feedback from the sensors, reference signal, and the MPC controller should all run at the same step size

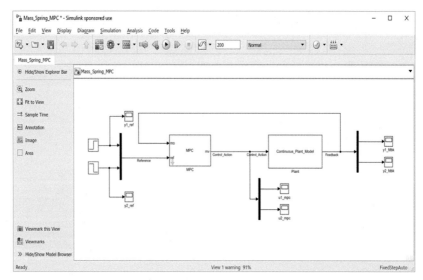

FIGURE 3.34 Restructured Simulink model for mass spring MPC controller tracking.

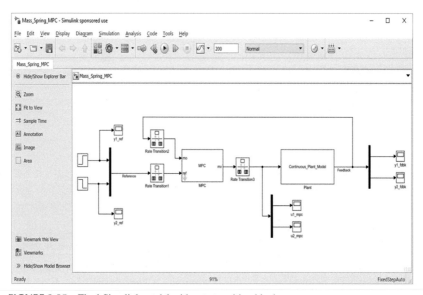

FIGURE 3.35 Final Simulink model with rate transition blocks.

when connected to the MPC block. Note that the rate transition block needs to be initialized for these three blocks. Initialize these blocks to [0;0]. The final model is saved as *Mass_Spring_MPC.slx* in Chapter_3/ Section_7. Fig. 3.35 shows the final Simulink model.

Chapter_3_Section_7_Script_1.m

```
%Book Title: Practical Design and Application of MPC
%Chapter: 3
%Section: 7
%Authors: Nassim Khaled and BibinPattel
%Last Modified: 10/15/2017
%% Load the discrete time mass damper system and run the system ID simulation
clc
clear all
bdcloseall
close all
% Load the Continuous Time Model of Mass Damper System
load Continuous_Plant_Model.mat
% Load the MPC Controller
load Mass_Spring_MPC.mat
% Open the model and simulate
open_system('Mass_Spring_MPC');
sim('Mass_Spring_MPC');
%% Plot the Controller Performance Tracking
subplot(221);
plot(u1_mpc(:,1),u1_mpc(:,2),'linewidth',2);
grid on
legend('u1 mpc')

subplot(223);
plot(u2_mpc(:,1),u2_mpc(:,2),'linewidth',2);
grid on
xlabel('Time [s]')
legend('u2 mpc')

subplot(222);
plot(y1_ref(:,1),y1_ref(:,2),'linewidth',2);
hold all
plot(y1_fdbk(:,1),y1_fdbk(:,2),'--','linewidth',2);
grid on
legend('y1 ref','y1 fdbk')

subplot(224);
plot(y2_ref(:,1),y2_ref(:,2),'linewidth',2);
hold all
plot(y2_fdbk(:,1),y2_fdbk(:,2),'--','linewidth',2);
grid on
xlabel('Time [s]')
legend('y2 ref','y2 fdbk')

set(gcf,'color',[1 1 1 ]);
grid on
```

Run the script *Chapter_3_Section_7_Script_1.m* to open the model, load the plant and controller, run the simulation, and plot the controller performance tracking. The controller tracking performance is plotted in Fig. 3.36. Some overshoot is observed at the beginning of the simulation. Fig. 3.37 shows a zoomed-in version of the figure. The main issue with the tuning of the controller is the prediction horizon of the MPC. The prediction horizon is 1 second (10 samples multiplied by 0.1 second). We will increase the

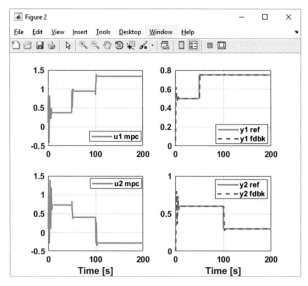

FIGURE 3.36 MPC controller performance.

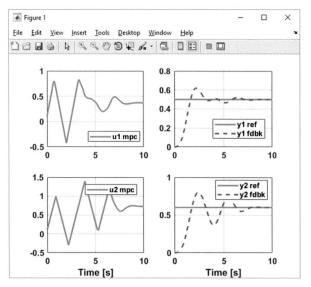

FIGURE 3.37 MPC controller performance for first 10 s.

prediction horizon to 100 and the control horizon to 20. Fig. 3.38 shows the improved performance once the prediction and control horizons are increased. To replicate the results of Fig. 3.38, run *Chapter_3_ Section_7_Script_2.m.* The corresponding *MPC Design Task* is saved under the name *MPC_DesignTask_Chapter_3_Section_7.mat.*

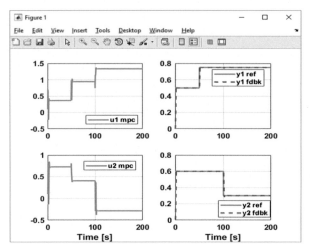

FIGURE 3.38   MPC controller performance for increased prediction and control horizons.

There are practical constraints to consider when deciding on the prediction and control horizon. Executing MPC with a prediction horizon of 100 and control horizon of 20 takes significant computational resources and is not typical. Another way to mitigate performance issues like the one encountered in Fig. 3.36 is to increase the step size of the controller from 0.1 to 0.2 seconds while keeping the prediction horizon at 10 and the control horizon to 2. This increases the prediction window of the MPC to 2 seconds instead of 1 second.

3.8   APPLICATION PROBLEM

Download the material for Chapter_3/Section 7. We will test the controller performance while varying the sample time, prediction, and control horizon.

1. Open MPC Designer App and load *MPC_DesignTask_Chapter_3_Section_7.mat*
2. Change the sample time to 0.4, prediction horizon to 25 and control horizon to 5.
3. Right click on the controller *Mass_Spring_MPC* on the left and save it (Export to Workspace).
4. Save the workspace into the file *Mass_Spring_MPC_Updated.mat*
5. Open *Mass_Spring_MPC.slx* and change the output port sample time of the rate transition of feedback and reference signals to 0.4 to match that of the MPC controller. Save the model and close it.
6. RUN *Chapter_3_Section_7_Script_2.m*. Is there a significant change in the performance of the controller?

The solution of the application problem can be found in the Chapter_3/ Section 8 folder. To check the final Simulink model, open *Mass_Spring_MPC.slx*. To generate the tracking performance, run *Chapter_3_Section_7_Script_2.m*.

REFERENCES

[1] S.J. Qin, T.A. Badgwell, A survey of industrial model predictive control technology, Control Eng. Practice 11 (7) (2003) 733−764.

[2] H. Zheng, R.R. Negenborn, G. Lodewijks, Trajectory tracking of autonomous vessels using model predictive control, in Proceedings of the 19th IFAC World Congress (IFAC WC'14), Cape Town, South Africa, pp. 8812−8818, August 2014.

[3] S. Oh, J. Sun, Path following of underactuated marine surface vessels using line-of-sight based model predictive control, Ocean Eng. 37 (2010) 289−295.

[4] N. Khaled, M. Cunningham, J. Pekar, A. Fuxman, O. Santin, Multivariable Control of Dual Loop EGR Diesel Engine with a Variable Geometry Turbo, SAE World Congress and Exhibition, 2014.

[5] E.F. Camacho, C. Bordons, Model Predictive Control, Springer-Verlag, London, 1999.

Chapter 4

System Identification for a Ship

4.1 INTRODUCTION

The demand for smart devices and artificial intelligence is on the rise. In addition to autonomous cars, there is great interest in autonomous ships. The next three chapters will handle, in detail, Model Predictive Control (MPC) design and robustness analyses for a ship. While Chapter 3, MPC Design of a Double-Mass Spring System, covered the overall process of system identification and controller design, this chapter will only cover system identification for a ship (Fig. 4.1). The controller design will be covered in the next chapter. This was intentionally done due to the complexity level of controlling a ship compared with controlling a double-mass spring system.

In Section 4.2, the nonlinear dynamic model for a ship will be presented.

Following the system identification process outlined in Chapter 3, MPC Design of a Double-Mass Spring System, Section 4.3 will walk the reader through the steps for obtaining the linear model for the ship using the MATLAB System Identification Toolbox.

In Section 4.4 the reader is challenged with an application problem that will help you to practice the methods discussed in this chapter.

All the codes used in the chapter can be downloaded from *MATLAB File Exchange*. Follow the link below and search for the ISBN or title of this book.

https://www.mathworks.com/matlabcentral/fileexchange/

Alternatively, the reader can also download the material and other resources from the dedicated website or contact the authors for further help.

https://www.practicalmpc.com/

FIGURE 4.1 System identification.

Practical Design and Application of Model Predictive Control.
DOI: https://doi.org/10.1016/B978-0-12-813918-9.00004-6

4.2 PLANT MODEL OF A SHIP

The dynamics of a ship are highly nonlinear [1−4]. Ships are significantly influenced by environmental disturbances generated by winds, random sea waves, and currents. Furthermore, a ship model is a multi-input multi-output system that would serve as a good test to design and test the MPC in later chapters.

The nonlinear model of the ship developed for the use with this book follows the work of Khaled and Chalhoub [1] closely. The model includes the six degrees of freedom of a 100 m-long ship. The two actuators of a ship are the single propeller and rudder (these are the manipulated variables). The measured outputs of the model are the forward speed of the ship (u), the lateral speed of the ship (v), turning rate around its z-axis (r) and the heading of the ship (ψ). Fig. 4.2 shows the block diagram of the full model that includes the unmeasured disturbances, manipulated variables and the measured outputs. Fig. 4.3 shows a schematic representation of a ship, its propeller, rudder, the global coordinate system, and the ship's local coordinate system.

For the reader to become familiar with the model of a ship with its inputs and outputs, a simulation was run where the manipulated variables were stepped according to the top two plots as shown in Fig. 4.4. The rudder angle, α, was stepped in the positive direction, similarly the speed of the propeller, n_{pr}, was stepped in the positive direction. Stepping the rudder in the positive direction forced the ship's heading (ψ) to turn in the negative direction. The turning rate of the ship (r) moved in the negative direction. This

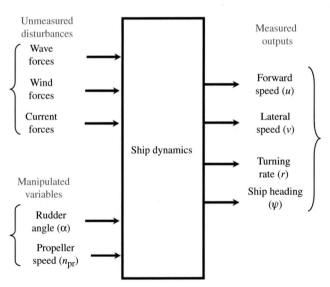

FIGURE 4.2 Block diagram of the full model of the ship.

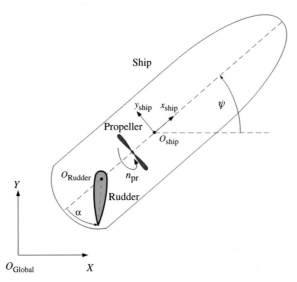

FIGURE 4.3 Schematic of the ship, propeller and rudder.

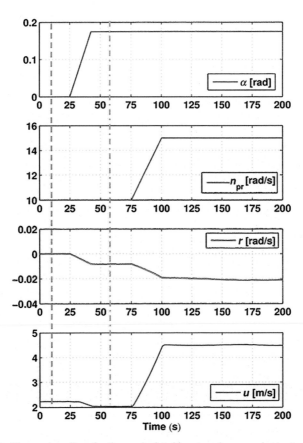

FIGURE 4.4 Time series values for the manipulated inputs and measured outputs of the ship.

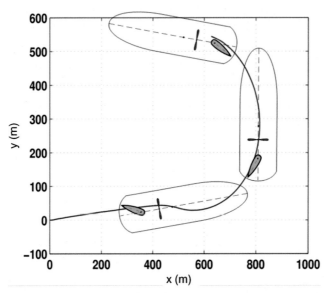

FIGURE 4.5 Change in ship position and heading due to a change in the rudder angle.

can be observed through the third plot from the top in Fig. 4.4. The fourth plot from the top of Fig. 4.4 shows that increasing the rotational speed of the propeller led to an increase in the ship's speed from an initial 2 m/s to about 4.5 m/s. To facilitate reading Fig. 4.4, a green dashed line was added indicating the time at which the rudder started to rotate in the positive direction. Additionally, an orange dash-dot line was added indicating the time at which the propeller speed was increased. It is worthwhile noting the evident coupling between the rudder angle, α, and the ship speed, u. The ship speed decreased as the ship started turning, due to the change in the rudder angle. Similarly, as the propeller speed increased, the magnitude of the turning rate increased.

To better visualize the outcome of the above simulation, Fig. 4.5 shows the x versus y position of the ship. Moreover, a sketch of the ship is overlaid on the plot to show both the heading of the ship as well as that of the rudder.

4.3 DATA-BASED LINEAR APPROXIMATION OF THE SHIP'S DYNAMICS

Using the plant model of the ship introduced in the previous section combined with the process outlined in Chapter 3, MPC Design of a Double-Mass Spring System, we demonstrate the process of system identification of a

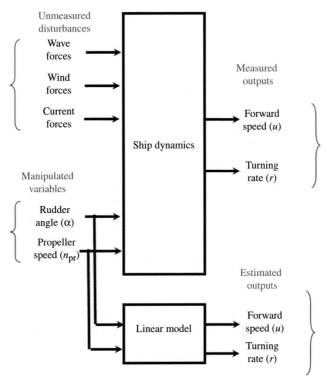

FIGURE 4.6   Block diagram of the model of the ship and its linear approximation.

ship. Instead of using all the measured outputs of the ship model outlined in Fig. 4.2, only the ship's forward speed, u, and ship's turning rate, r, will be involved as outputs in the system identification. These two outputs will be controlled by MPC in the next chapter. Both the rudder angle, α, and the propeller speed, n, will be used as manipulated inputs (Fig. 4.6). The ship is operating under the influence of unmeasured waves, wind and current disturbances. Since these external factors are not measured, they will contribute to the inaccuracy of the identified model. When doing data collection on a real ship, it is better to choose a day with calm waters and low wind speeds.

Fig. 4.7 shows the outcome of the system identification that was obtained using the MATLAB System Identification Toolbox. The figure shows that the linear model can predict the rate of the turning of the ship, as well as the forward speed, reasonably well. To quantify how well the identified model fits the data, we will use the metric goodness of fit (*gof*). It is a number from 0% to 100%. 0 means a bad fit while 100 means an excellent one. We compute *gof* using the *goodnessOfFit* function in *MATLAB* with the *NRMSE* (Normalized Root Mean Square Error) option.

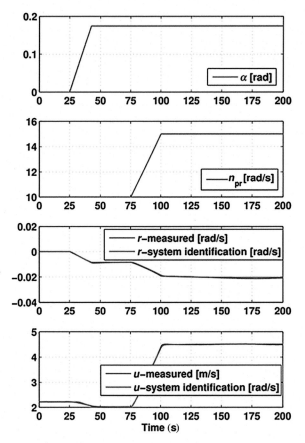

FIGURE 4.7 System identification results for the ship.

We define *gof* as:

$$\text{gof} = 100 \times \left(1 - \frac{\left\| y_{\text{measured}} - y_{\text{model}} \right\|}{\left\| y_{\text{measured}} - \text{mean}(y_{\text{model}}) \right\|} \right) \quad (4.1)$$

where $\| \|$ is the second norm of a vector.

The *gof* for Fig. 4.7 are 98.5% and 93.2% for the turning rate and forward speed of the ship respectively.

Below are some useful guidelines to obtain a good fit:

- If using a simulation model, discard the first few samples as most dynamic models need a few iterations before converging to the solution, especially when the states of the model are not properly initialized

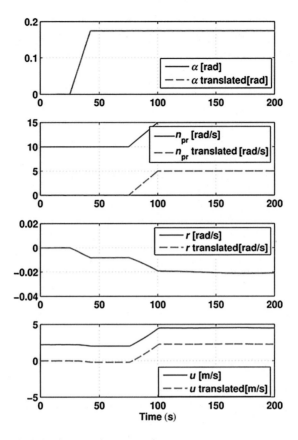

FIGURE 4.8 Translating inputs and outputs at time zero to zero.

- Try to conduct the data collection where the unmeasured disturbance is kept minimal. Otherwise the system identification will be misled by noise.
- Translate the measured outputs of the system as well as the manipulated variables to 0. The offsets can be added back once the system identification is concluded. Fig. 4.8 shows the translated measured outputs and the manipulated variables. This step doesn't improve the system identification result, but is a good practice to follow when preparing to design a MPC controller.
- Use *n4sid* function in MATLAB to generate a preliminary linear model and then feed the resulting model to the *pem* function.
- Use *goodnessOfFit* function in MATLAB to calculate *gof*.

- If the gof is less than 75% in the case of minimal external disturbance, then the system identification didn't succeed. In such a case, the system identification algorithm might have failed to find a good fitting model or the input/output data is not of a good quality.

The following M-script file, *Chapter_4_Section_3.m*, simulates the ship dynamics when subjected to the manipulated variables as prescribed in Fig. 4.6. Some comments have been added to briefly explain the M-script file. The inputs and outputs at time zero are translated to zero (Fig. 4.8). This is a standard practice that we follow in the book but system identification could have been done without translation. The M-script can be found in the Chapter 4/ Section_3 folder of the extra material of this book. The folder can be downloaded from Elsevier's webpage: https://www.elsevier.com/books-and-journals/book-companion/9780128139189.

Chapter_4_Section_3.m

```
%Book Title: Practical Design and Application of MPC
%Chapter: 4
%Section: 3
%Title: System Identification for a Ship
%Authors: Nassim Khaled and Bibin Pattel
%Last Modified: 4/24/2017

%% Plotting Settings
set(0,'DefaultLineLineWidth',1)
set(0,'DefaultAxesFontWeight','bold')
set(0,'DefaultAxesFontName','Arial')
set(0,'DefaultAxesFontSize',12)

%% Simulate the model
sim('Chapter_4_Section_3_Model.slx') %Simulates the model
%% Plot the propeller speed
figure
plot(npr_user_provided(:,1),npr_user_provided(:,2))
hold on
plot(npr_achievable(:,1),npr_achievable(:,2))
legend('n_p_r','n_p_r achievable')
xlabel('Time [s]')
title('Propeller Speed [radians/s]')
%% Plot the rudder angle
figure
plot(alpha_user_provided (:,1),alpha_user_provided(:,2))
hold on
plot(alpha_achievable(:,1),alpha_achievable(:,2))
legend('alpha user provided','alpha achievabled')
xlabel('Time [s]')
title('Rudder Angle [radians]')
%% Grouping simulation data and neglecting first few seconds(Nstart*stp_sz)
Nstart = 100; %Clip the data starting from Nstart as the first few seconds of the
simulation should be discarded

Measured_Outputs=[r_simulation(Nstart:end,2) u_simulation(Nstart:end,2)]; %Measured
outputs
Manipipulated_Variables=[alpha_achievable(Nstart:end,2) npr_achievable(Nstart:end,2)];
%Manipulated variables

%% Obtaining initial conditions at step time=Nstart
Measured_Outputs_Nstart=Measured_Outputs(Nstart,:); %Capturing the measured outputs at
step time= Nstart
Manipipulated_Variables_Nstart=Manipipulated_Variables(Nstart,:); %Capturing the
manipulated variables at step time= Nstart

%% Forcing response to start from zero initial conditions
Measured_Outputs_zero_initial_conditions=Measured_Outputs-
repmat(Measured_Outputs_Nstart,length(Measured_Outputs),1); %Subtracting initial
conditions for measured outputs to obtain zero response at step time= Nstart
Manipipulated_Variables_zero_initial_conditions=Manipipulated_Variables-
repmat(Manipipulated_Variables_Nstart,length(Manipipulated_Variables),1); %Subtracting
initial conditions for manipulated variables to obtain zero actuation at step time=
Nstart

%% Prepare date for system identification
stp_sz=1;
data=iddata(Measured_Outputs_zero_initial_conditions(1:20:end,:),Manipipulated_Variables
_zero_initial_conditions(1:20:end,:),stp_sz); %data is packaged for system
identification using iddata

%% Generate a preliminary 4th order system that fits the data
sys1=n4sid(data,2,'Form','canonical','DisturbanceModel','None','InputDelay',[0
0]','InitialState','zero'); %n4sid generates a preliminary system in the canonical form
%with zero disturbance, zero delay and zero initial conditions
%% Generate a more refined system
sys2 = pem(data,sys1,'InitialState','zero') %pem generates a more refined system that
fits the data better

%% Define the options for comparing the various identified systems
opt = compareOptions('InitialCondition','Z');
[Y,fit,x0]=compare(data,sys2);
```

```
Y_1=Y.OutputData(:,1);
Y_2=Y.OutputData(:,2);

%% Plotting the system identification results
figure
% figure('Position',[358 0 500 1100])
subplot(4,1,1)
plot(0:stp_sz:stp_sz*(length(Y_1)-
1),Manipipulated_Variables_zero_initial_conditions(1:20:end,1))
grid on
legend('\alpha [rad]')
subplot(4,1,2)
plot(0:stp_sz:stp_sz*(length(Y_1)-
1),Manipipulated_Variables_zero_initial_conditions(1:20:end,2))
grid on
legend('n_p_r [rad/s]')
subplot(4,1,3)
plot(0:stp_sz:stp_sz*(length(Y_1)-
1),Measured_Outputs_zero_initial_conditions(1:20:end,1))
grid on
hold on
plot(0:stp_sz:stp_sz*(length(Y_1) 1),Y_1,'r')
legend('r-measured [rad/s]','r-system identification [rad/s]')
subplot(4,1,4)
plot(0:stp_sz:stp_sz*(length(Y_1)-
1),Measured_Outputs_zero_initial_conditions(1:20:end,2))
grid on
hold on
plot(0:stp_sz:stp_sz*(length(Y_1)-1),Y_2,'r')
legend('u-measured [m/s]','u-system identification [m/s]')
xlabel('Time [s]')

%% Compute the goodness of fit [%]
gof_r=goodnessOfFit(Measured_Outputs_zero_initial_conditions(1:20:end,2),Y_2,'NRMSE')*10
0 % gof for r
gof_u=goodnessOfFit(Measured_Outputs_zero_initial_conditions(1:20:end,1),Y_1,'NRMSE')*10
0 % gof for u

%% Assess stability of the identified system
pole(sys2) %poles of sys2
isstable(sys2)
```

4.4 APPLICATION PROBLEM: SYSTEM IDENTIFICATION OF SHIP DYNAMICS

In this application, the same plant model for the ship discussed in Section 4.2 will be used.

1. Run the Simulink model (*Chapter_4_Section_4_Model.slx*) provided in the material for Chapter 4/Section_4. The simulation parameters will be time-based arrays that are sent to the workspace by the simulation. Below is a table of the parameter names and their description:

Simulation Parameter Name	Description	Unit
u_simulation	Forward speed of the ship	m/s
psi_simulation	Heading of the ship	Radians
r_simulation	Turning rate of the ship	Radians/s
npr_user_provided	User commanded propeller speed	Radians/s
npr_achievable	Actual propeller speed	Radians/s
alpha_user_provided	User commanded rudder angle	Radians
alpha_achievable	Actual rudder angle	Radians

2. Plot *npr_user_provided* and *npr_achievable* on the same plot as a function of time. Notice that the provided command wasn't achieved due to the dynamics of the propeller.
3. Plot *alpha_user_provided* and *alpha_achievable* on the same plot as a function of time. Notice that the provided command wasn't achieved due to the dynamics of the propeller.
4. Using the System Identification Toolbox in MATLAB, compute a discrete state space linear approximation of the ship dynamics. Use *alpha_achievable* and *npr_achievable* as the manipulated variables and *u_simulation* and *r_simulation* as the measured outputs. Hint: Clip the first 100 samples of the simulation data. Also, force the measured and manipulated variables to start from zero by subtracting the offset.
5. Calculate the normalized root mean square error for the developed model.
6. Is the obtained linear system stable?

For a complete solution of the problem, the reader can read and run *Chapter_4_Section_4_Script.m* that can be found in the Chapter 4/Section_4 folder.

REFERENCES

[1] N. Khaled, N.G. Chalhoub, A Dynamic model and a robust controller for a fully actuated marine surface vessel, J. Vibrat. Control 17 (6) (2010) 801−812.
[2] T.I. Fossen, Guidance and Control of Ocean Marine Vehicles, John Wiley and Sons Ltd, New York, 1994.
[3] T. Perez, Ship Motion Control, Springer Verlag, London, 2005.
[4] J.N. Newman, Marine Hydrodynamics, MIT Press, Cambridge, Massachusetts, 1977.

Chapter 5

Single MPC Design for a Ship

5.1 INTRODUCTION

This chapter guides the reader through the process of designing a linear MPC controller for a ship (Fig. 5.1). The turning rate and ship speed are controlled using the rudder and propeller. These actuators operate under physical constraints that will be highlighted later. MPC can be a very effective and practical controller for such a multi-input multi-output (MIMO) system with constraints.

The performance of linear MPC is dependent on the envelope of operation of the controller (range of operation of the measured and unmeasured disturbance, manipulated variables, sensing, and actuation system errors) as well as the nonlinearity of the plant. Applying the designed controller outside the envelope of operation can lead to undesired performance of the system and instability.

This chapter assumes that the reader is familiar with MPC. If this is not the case, the reader can go back to Chapters 2 and 3 to get acquainted with this model-based control strategy. This chapter also assumes the existence of a linear model for the nonlinear plant it is recommended that readers go over the previous chapter to understand the plant model as well as the system identification process.

The chapter will walk readers through the detailed process of designing a MPC controller for a ship. Tuning challenges and hurdles that might face the designer will be outlined along with recommended solutions.

The chapter concludes with an application problem that challenges the reader to implement the methods discussed.

All the codes used in the chapter can be downloaded from *MATLAB File Exchange*. Follow the link below and search for the ISBN or title of the book.

https://www.mathworks.com/matlabcentral/fileexchange/

Alternatively, the reader can also download the material and other resources from the dedicated website or contact the authors for further help.

https://www.practicalmpc.com/

Practical Design and Application of Model Predictive Control.
DOI: https://doi.org/10.1016/B978-0-12-813918-9.00005-8

FIGURE 5.1 Controller design.

5.2 UNDERSTANDING THE REQUIREMENTS FOR THE CONTROLLER

In industrial applications, the controller being developed is usually a part of a product. Product development usually follows the V-diagram (Fig. 5.2) [1,2]. Thus, as the first step, the control designer needs to understand the requirements. The challenge is that defining control requirements is often not straightforward they are hard to link to the system's performance and are poorly written. The reason is that control requirements are hard to define independently from the capability of the plant, actuation system, and the sensing system. Therefore, in many cases, the control designer is assigned the task of defining the requirements instead of being handed the requirements that need to be met. The designer needs to agree with his/her customer on the requirements, the method for verifying and validating these requirements, conditions under which the requirements will be verified and validated, as well as the specifications of the sensors and actuators.

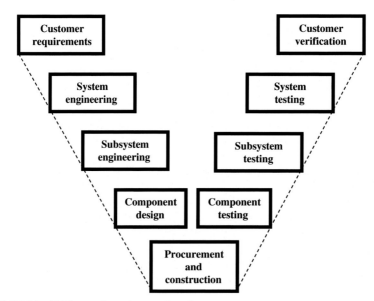

FIGURE 5.2 V-Diagram in systems engineering.

It is an essential part of a control engineer to evaluate whether the requirements are within the physical hardware capability. Running open-loop testing for the plant in hand, as well as understanding the actuators' dynamic response in addition to understanding the sensors specifications are all within the control engineer's scope of work.

5.3   REQUIREMENTS FOR THE SHIP CONTROLLER

In line with what was introduced in Chapter 4, the control design problem is that of ship navigation. The objective of the controller is to control the ship's forward speed and turning rate. The actuators available are the rudder machine and propeller. The validation and verification set points are provided by the customer as a time domain reference signal for both the ship's forward speed and turning rate.

The requirements are as follows:

Requirement #1:
The maximum absolute error for the turning rate of the ship during the provided maneuver shall not exceed 4.5×10^{-3} rad/s.

Requirement #2:
The maximum absolute error for the ship's forward speed during the provided maneuver shall not exceed 0.42 m/s.

Requirement #3:
The root mean square error for the turning rate of the ship during the provided maneuver shall not exceed 8×10^{-4} rad/s.

Requirement #4:
The root mean square error for the ship forward speed during the provided maneuver shall not exceed 0.25 m/s.

Requirement #5:
The forward speed during the provided maneuver shall not exceed the reference forward speed by more than 0.5 m/s.

Fig. 5.3 shows the time series maneuver that will be used to check whether the control design and tuning meets requirements 1 through to 5. From Fig. 5.3, the ship's turning rate at time zero $r = 0$ rad/s and the ship's forward speed is $u = 2.22$ m/s. The actuators' position for this maneuver are $\alpha = 0$ rad and $n = 10$ rad/s for the rudder and propeller respectively. These positions for the actuators can also be referred to as the *feed forward* or *nominal* values.

Fig. 5.4 shows the Simulink model for the ship in addition to the reference generator that will be used for verification and validation of the designed MPC. The Simulink model, **Chapter_5_Model_and_Reference.slx**, can be found in **Chapter 5/Section_3** folder that can be downloaded from https://www.elsevier.com/books-and-journals/book-companion/9780128139189.

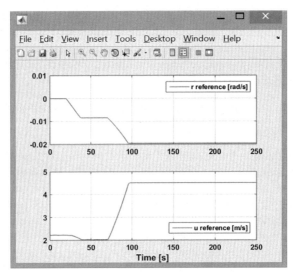

FIGURE 5.3 Validation and verification setpoints.

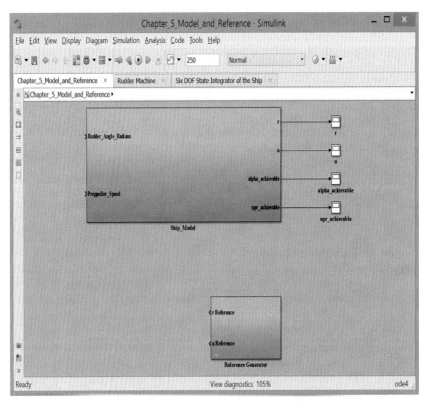

FIGURE 5.4 Simulink model of the ship and the reference generator for the maneuver.

5.4 PHYSICAL CONSTRAINTS OF THE SHIP

Physical properties of the system were obtained from open-loop testing for the ship. These will greatly enhance the performance of MPC.

The rudder angle operates between −0.39 to 0.39 rad. The range for the rudder (which is the difference between the maximum and minimum values) is 0.78 rad. As for the propeller, the speed can take any value between 0.1 and 30 rad/s. The range for the propeller is rounded to 30 rad/s. The rate of the rudder can take any value between −0.01 and 0.01 rad/s, whereas the rate of the propeller can take any value between − 0.2 and 0.2 m/s^2.

As for the ship's turning rate, it operates between −0.02 and 0.02 rad/s. Thus, the range is 0.04 rad/s. As for the ship's forward speed, it can't exceed 25 m/s. So the range of speed will be set to 25 m/s.

Providing a MPC with the constraints for the actuators, their rates, and the range for the outputs allows the quadratic solver of the MPC to choose values that are realistic and reduces tuning time.

5.5 HANDLING CONSTRAINTS IN MPC

One of the main advantages of MPC is the ability to handle constraints. Hard constraints must be respected by the solution computed by the quadratic solver of MPC. A mathematical solution that satisfies the hard constraints, the manipulated variables, will be set to their previous values. Persistence of such conditions leads to a loss of control.

We can always set hard constraints on the manipulated variables without causing infeasible solutions. Similarly, we can set hard constraints on the rate of the manipulated variables. The issue of infeasible solutions might arise if we set hard constraints on both the manipulated variables and their rates at the same time. To handle such potential problems, we can "soften" the constraints. The quadratic solver of MPC, if it computes an infeasible solution, will sacrifice the softened constraints to find a feasible solution.

Since we are going to set constraints on the manipulated variables as well as their rates, we need to soften at least one of these two sets. Fig. 5.5 shows constraint softening settings in *mpcDesigner* which can be found in the Tuning tab under the Constraints button.

To soften a constraint, set the corresponding Equal Concern for the Relaxation (ECR) value to a positive scalar. To soften the constraint more, increase the ECR value.

In the case of the ship, we will soften the rate of the manipulated variables and keep the constraints of the manipulated variables as hard constraints. The output constraints will be left as default soft constraints (which means the ECR minimum and maximum values are set to 1).

For more information, the user is referred to the help documentation of *mpcDesigner*.

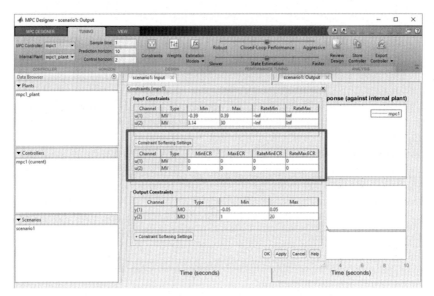

FIGURE 5.5 Constraint softening settings.

5.6 DESIGNING A MPC CONTROLLER FOR THE SHIP USING MATLAB

The process of designing MPC in MATLAB is well-packaged in the **MPC Designer App**. Download the **Chapter_5/Section_6** material and make the folder the working directory. Similarly to Chapter 3, follow the steps below to design the MPC controller for the ship.

1. To start the design process, download and load the linear model, *Chapter_5_Section_6_Linear_Model.mat*, for the ship that was prepared for this chapter.
2. In the MATLAB command window, type *mpcDesigner* to open *MPC Designer App* GUI as shown in Fig. 5.6.
3. Click on the *MPC Structure* button to configure the controller (Fig. 5.7).
4. Select the plant model (*ship_linear_model* in Fig. 5.8). You can click on the model for a preview of the structure (Fig. 5.9).
5. Close the preview of the linear model and click *Define and Import* (Fig. 5.10). You might need to wait for a minute until the default MPC controller is generated. Fig. 5.11 shows how the MPC default setup should look like when all the previous steps have been executed properly.

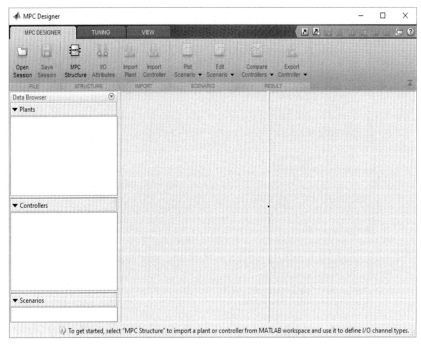

FIGURE 5.6 *MPC Designer App* GUI.

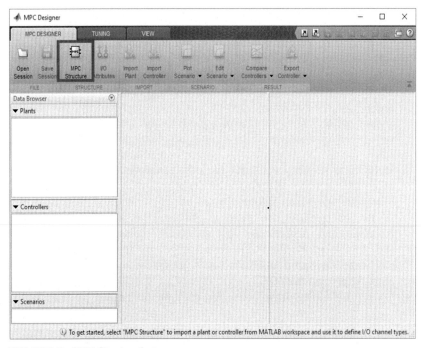

FIGURE 5.7 *MPC Structure* button.

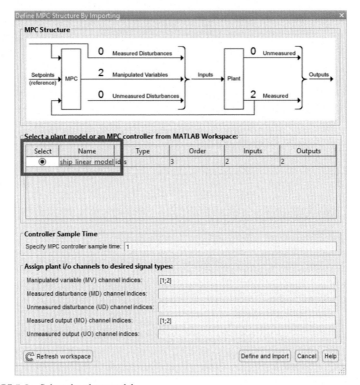

FIGURE 5.8   Select the plant model.

6. Click the **I/O Attributes** button to define the names, units, scaling factors, and nominal values of the inputs and outputs of the controller (Fig. 5.12).

7. Define the units of the inputs and outputs as indicated in Fig. 5.13. **u1** and **u2** are the rudder and propeller actuators respectively, whereas **y1** and **y2** are the ship's turning rate and forward speed respectively. This step is optional as the units are not necessary for the design of the MPC.

8. In order to properly construct the cost function, some work needs to take place in order to scale the actuation effort as well as the outputs. Using the open-loop information about the inputs and outputs of the system provided in the previous section, the user can fill in the **Nominal** and **ScaleFactor** for the MPC controller as shown in Fig. 5.14. **ScaleFactor** is the set of ranges for the inputs and outputs that were introduced in section 5.4. After the values have been entered, click **Apply** and then **OK**.

9. To start the tuning of the controller, click on the **Tuning** tab (Fig. 5.15).

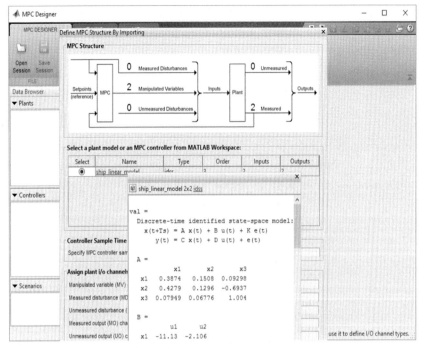

FIGURE 5.9 Linear model of the ship.

10. To choose the control and prediction horizon for MPC, it is important to know the time constant as well as the sampling frequency of the sensors. The time constant of the ship is in the order of 20 seconds and can be easily obtained by running step changes on the rudder and propeller, and computing the slower time constants between heading and turning rate. The prediction horizon selected here was half that of the time constant (10 seconds) of the system. If the prediction horizon is much smaller than the time constant of the system, there will be a noticeable degradation in the performance of the MPC. The sensors sampling frequency is the same as the control action execution frequency, which is 1 second. This means that the sensors are reporting back a reading on the ship's turning rate and speed every second and that the controller will command the actuators with a new position every second as well. The control horizon was chosen to be 2 seconds. Before setting these values in the tool, it is worthwhile noting that the *Control interval* is the function of the sample time for the discrete model for the MPC. In this case, the sample time was 1 second for the ship's linear model. By setting the *Control Interval* to 2 means that the *Control interval* = 1x2 = 2 seconds. Set the *Prediction horizon* and *Control horizon* to 10 and 2 respectively (Fig. 5.16).

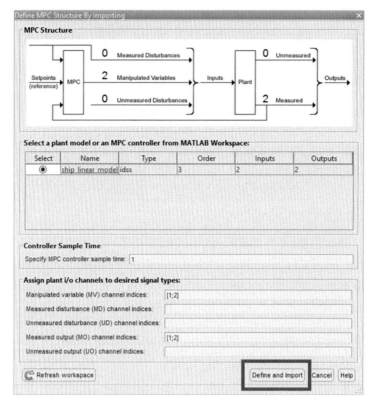

FIGURE 5.10 *Define and Import* button.

11. Click on the ***Constraints*** tab to define the range of the actuators, their rates as well as the range of the outputs (Fig. 5.17).
12. While executing, MPC will select values for the actuators that obey the constraints defined by the user. The constraints are the actuators minimum/maximum, actuators minimum/maximum rates, as well as the minimum/maximum for outputs. Using the physical constraints from section 5.5, input the minimum and maximum values. Fig. 5.18 shows the final setup of the constraints. Press **Apply** and then **OK**. You might need to wait for a minute for these changes to apply.
13. To avoid infeasible solutions as discussed in section 5.5, we will soften the constraints on the rate of the manipulated variables. The RateMinECR for both rudder rate and propeller will be set to zero (hard constraint) while RateMaxECR will be set at 1 (soft constraint). This provides the solver with the flexibility to relax the rate of the manipulated variables in order to obtain a feasible solution. Fig. 5.19 shows the setup for the ECR values. Press **Apply** and then **OK**.
14. Click on the ***Weight*** button to tune the MPC (Fig. 5.20).

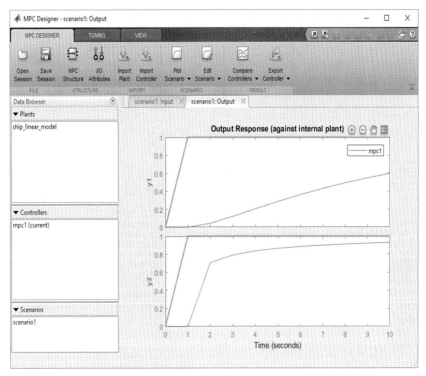

FIGURE 5.11 Default MPC controller.

15. In this application, we are not concerned with keeping the actuators around their nominal values. We don't want to include the actuators to the cost function. The *Input Weights* are set to 0. Since our prime objective is to force the outputs to track the setpoints, we will penalize the turning rate and ship speed equally. The *Output Weights* are set to 1. Keep the *Input Rate Weights* to 0.3. This will penalize continuous change on the command of the rudder and propeller. In the current tuning, there is more emphasis on tracking the input setpoints compared to penalizing the rate of the actuators (about three times more). Fig. 5.21 shows the final setup. Click **Apply** and **OK.**

16. Go back to MPC Designer tab (Fig. 5.22). The tracking is not good per the *Outputs* plots. There might be multiple reasons for that. It could be that the setpoints are unachievable and the actuators are hitting their constraints. It could also be that the tuning is not appropriate. Another possibility might be that the simulation duration is too short.

17. Click on *Edit Scenarios > scenario1* to set up more realistic simulation conditions for the MPC design (Fig. 5.23).

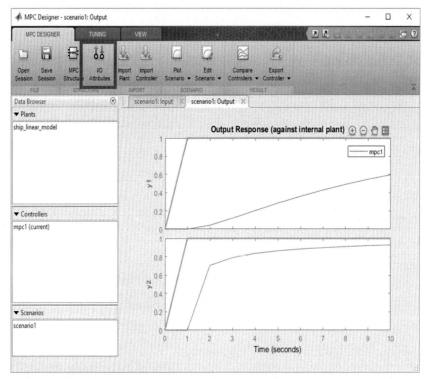

FIGURE 5.12 *I/O Attributes* button.

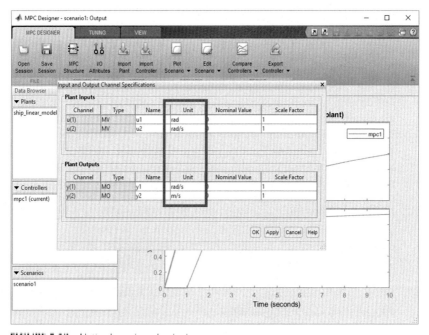

FIGURE 5.13 Units of inputs and outputs.

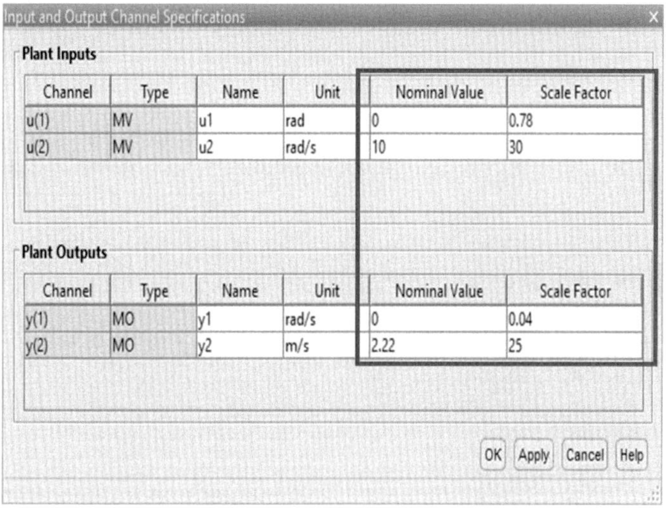

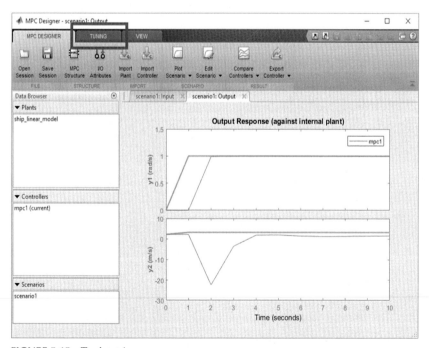

FIGURE 5.14   Nominal and **ScaleFactor** values for Inputs and Outputs.

FIGURE 5.15   Tuning tab.

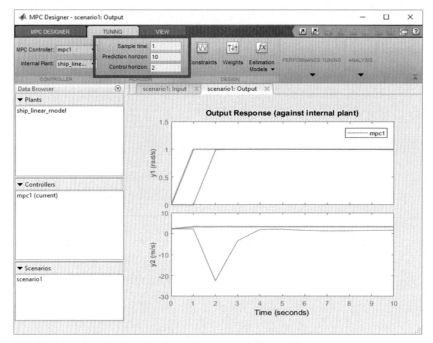

FIGURE 5.16   Sample Time, Prediction Horizon and Control Horizon.

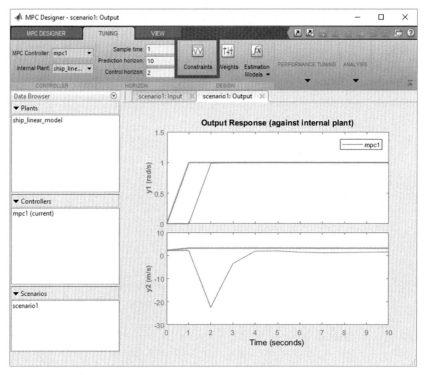

FIGURE 5.17   **Constraints** button.

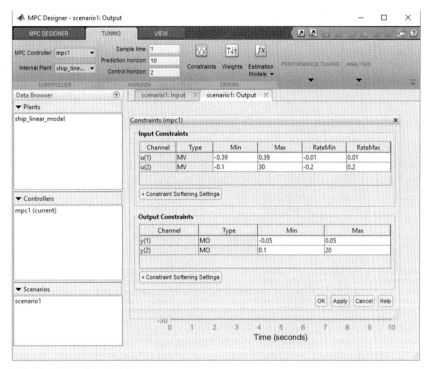

FIGURE 5.18 **Constraints** setup.

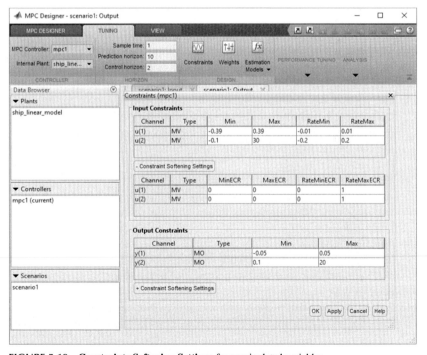

FIGURE 5.19 **Constraints Softening Settings** for manipulated variables.

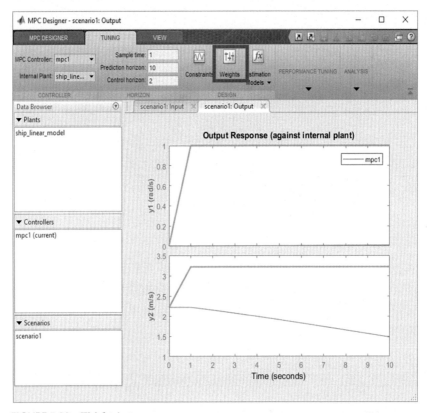

FIGURE 5.20 *Weights* button.

18. We will set the simulation time to be 100 s. The reference for the controller is selected to be a step (under *Type*) for both the ship's turning rate(0.01 rad/s) as well as the ship's forward speed 1.78 m/s) starting at 30 and 50 seconds respectively. The steps are increments around the nominal position. (Fig. 5.24). Click **Apply** and **OK.**
19. Fig. 5.25 shows the good tracking response of MPC. Fig. 5.26 shows the response of the propeller and rudder. Note that the MPC design tool uses the linear ship model combined with the MPC controller to simulate the design.
20. The final step before we can export the MPC controller, is to check the health of the MPC design. This is done by clicking on the tuning tab and then clicking on **Analysis > Review Design** (Fig. 5.27). You might need to wait for a minute before the report is generated.
21. Fig. 5.28 shows that there is one warning in orange related to *Hard MV Constraints*. Click on the hyperlinked text to see the recommended action.

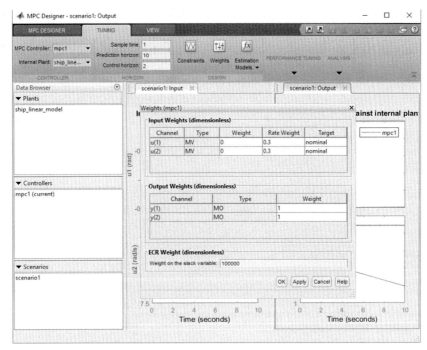

FIGURE 5.21 *Weights* setup.

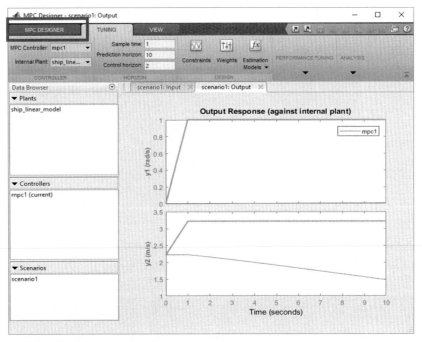

FIGURE 5.22 *MPC Designer* tab.

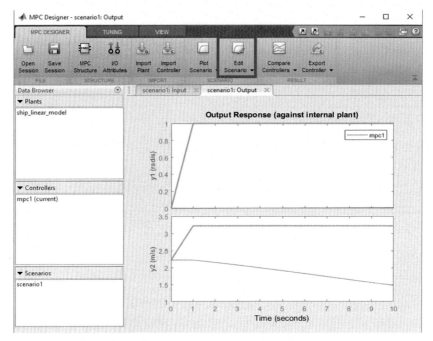

FIGURE 5.23 *Edit Scenario* button.

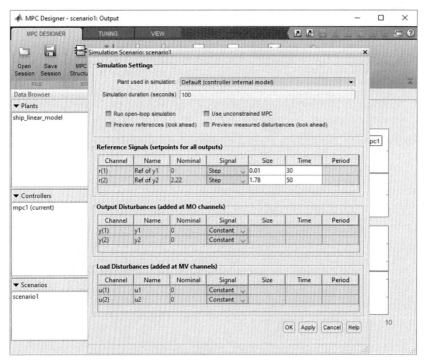

FIGURE 5.24 Simulation settings for *scenario1*.

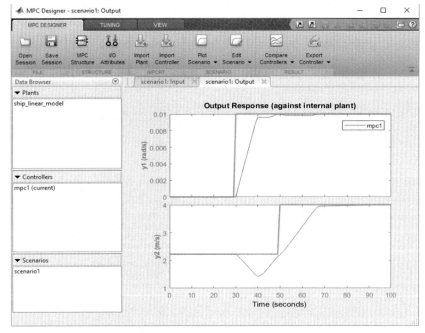

FIGURE 5.25   Output response of MPC.

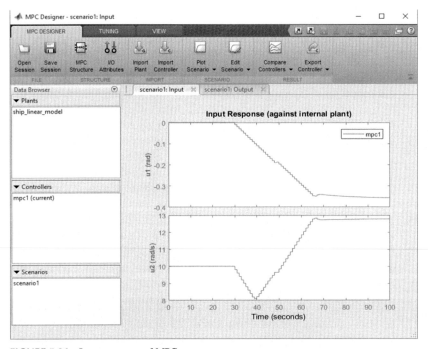

FIGURE 5.26   Input response of MPC.

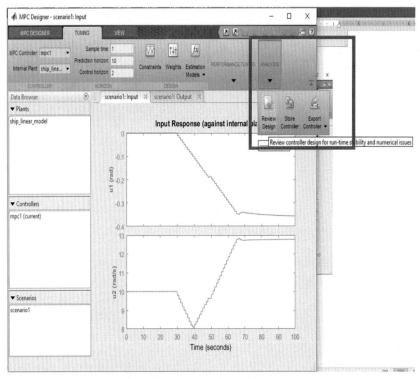

FIGURE 5.27 **Analysis > Review Design**.

22. Fig. 5.29 shows that the maximum value of the manipulated variable and the minimum rate which might conflict. This is applicable for both the rudder and the propeller. To resolve this warning, go back to the Constraints Softening Settings for the manipulated variables, and set RateMinECR for both rudder rate and propeller to a small non-zero value, say 0.01 (Fig. 5.30). Similarly set RateMaxECR to 0.1(Fig. 5.30).
23. Check the health of the MPC design by clicking on the tuning tab. Fig. 5.31 shows that the designed MPC passed all the checks.
24. Click the **Export Controller** button and select **Export Controller** (Fig. 5.32).
25. Tick beside mpc1 and click Export (Fig. 5.33). MATLAB will export **mpc1** into the workspace.
26. It is a good practice to save the session in case you need to go back and redesign or retune the MPC. Click on **Save Session** under MPC Designer and save the session (Fig. 5.34). MATLAB will save the session in a **.mat** format which can be loaded in future sessions.

The above session can be downloaded and loaded for verification. It is saved under ***MPCDesignTask_Chapter_5_Section_6.mat***

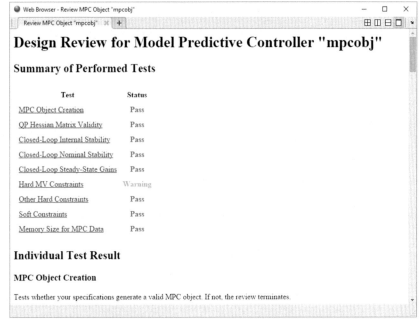

FIGURE 5.28   **MPC design** report.

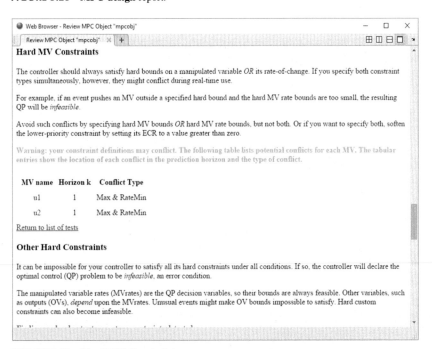

FIGURE 5.29   **Hard MV Constraints**.

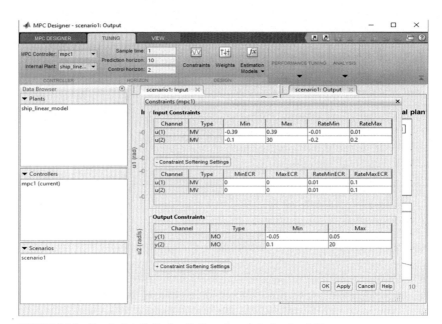

FIGURE 5.30 **Constraints Softening Settings** for manipulated variables.

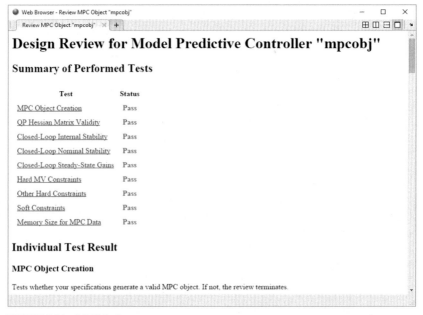

FIGURE 5.31 **MPC design** report.

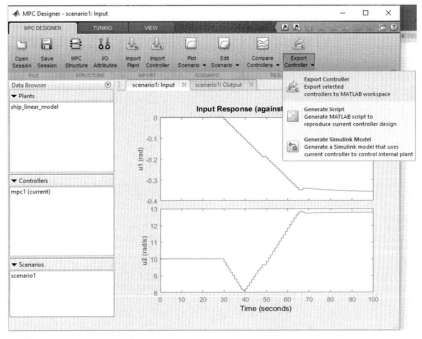

FIGURE 5.32 **Export Controller** button.

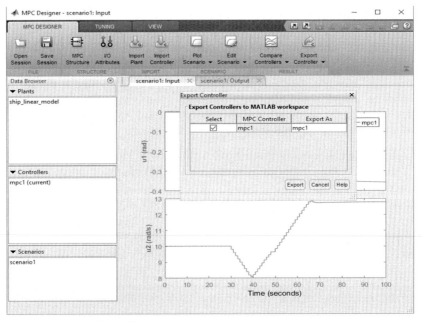

FIGURE 5.33 **mpc1** controller.

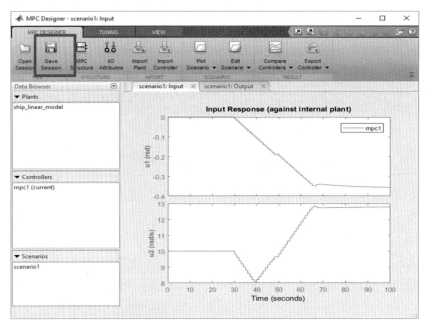

FIGURE 5.34   **Save Session** button.

5.7   INTEGRATING MPC WITH SIMULINK MODEL

In this section, the designed MPC will be integrated with Simulink for test-ing. To do so, the model *Chapter_5_Model_and_Reference.slx* (Fig. 5.4) will be used as a test platform. Download Chapter 5/Section_7 material and set it as the working directory in MATLAB as the model will load the necessary *.mat* file. In case you were not able to properly design the MPC controller from the previous section, you will need to load *MPC_Chapter_5_Section_7.mat*. Follow the below steps to integrate the controller with the ship model:

1. Add a *MPC Controller* block into the **Chapter_5_Model_and_Reference. slx** model. The block can be found in *Model Predictive Control Toolbox* (Fig. 5.35).
2. Double click the *MPC Controller* block. Type *mpc1* in the parameter *MPC Controller*. Since there is no unmeasured disturbance in the model, untick the *Measured disturbance (md)* (Fig. 5.36). Click *OK* when done.
3. Rotate the MPC controller block by selecting it and pressing *Ctrl + R* twice on the keyboard.
4. Add two signal multiplexers. Rotate the first multiplexer by selecting it and pressing *Ctrl + R* twice on the keyboard. Repeat for the second multiplexer.

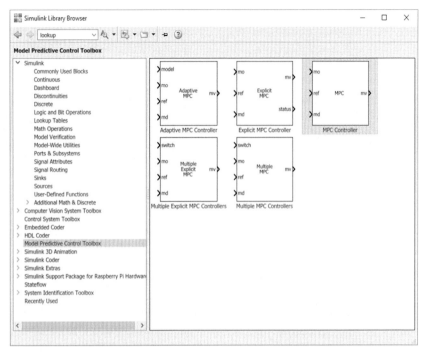

FIGURE 5.35   *MPC Controller* block.

5. Connect the measured *r* and *u* into the first multiplexer. Connect the reference *r* and *u* with the other multiplexer (Fig. 5.37).
6. Connect the multiplexers into the **MPC Controller** block as shown in Fig. 5.38.
7. Add a de-multiplexer to split the controller output into individual signals, rotate twice and connect to the ship model (Fig. 5.39).
8. One critical note before running the simulation is to understand the execution time needed for the controller and for the model. The Simulink model runs at a step size of 0.05 seconds. The controller doesn't need to execute at the same rate, especially since the dynamics of the ship are in orders of seconds. The linear model and the MPC controller have a step size of 1 second because the controller was designed to run at 1 second. Add three **Rate Transition** blocks (which can be found under **Simulink/ Signal Attributes**). The first block should be added before the *mo* port of the MPC block and its initial values should be set to [0;0]. The second block should be added before the *ref* port of the MPC block and its initial values should be set to [0;0]. As for the third block, it should be added to the *mv* port of the MPC block and its initial values should be set to [0;10] (initial angle of the rudder is 0 rad while the initial propeller speed is 10 rpm). Fig. 5.40 shows the final model with the **Rate Transition**

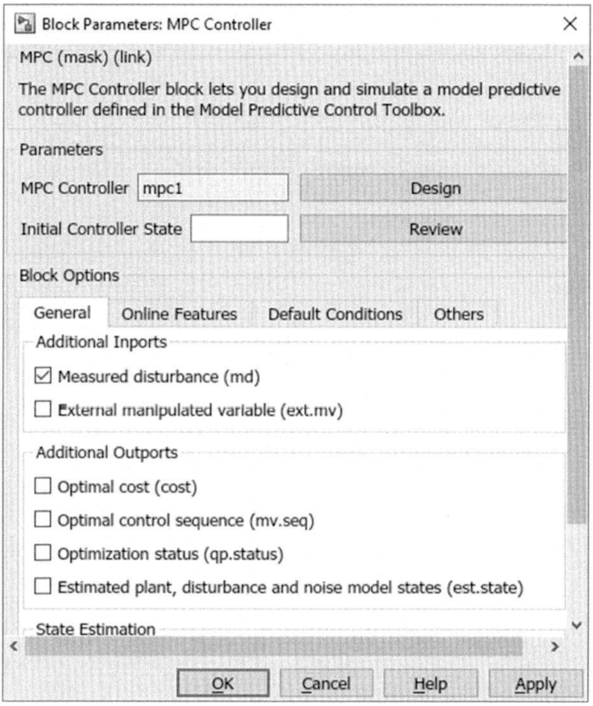

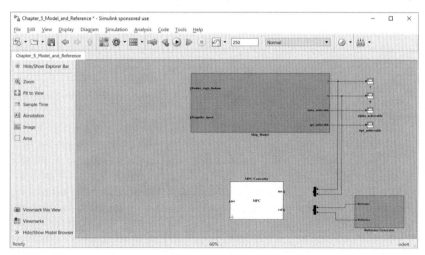

FIGURE 5.36 Embedding the controller in *MPC Controller* block edit.

FIGURE 5.37 Connecting measurements and reference to the multiplexers.

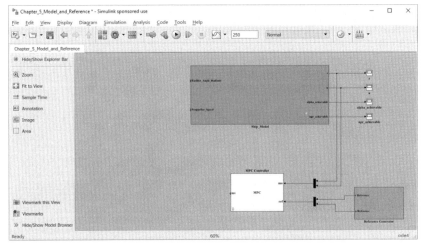

FIGURE 5.38   Connect the two multiplexers to the *MPC Controller.*

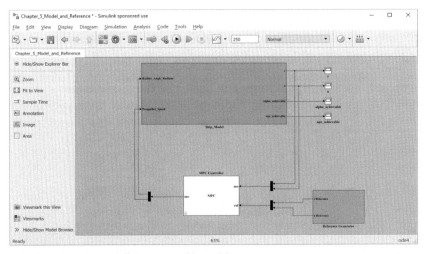

FIGURE 5.39   Connect MPC output to ship model.

blocks. Note that the ***Output port sample time*** of the **Rate Transition** blocks was left as **-1**. This means that Simulink will automatically detect the proper rate of the output.

9. Run the simulation and check tracking results (Fig. 5.41) and manipulated variables (Fig. 5.42). In Fig. 5.43, we zoom in on the manipulated variables. Notice how the command from MPC, n_{pr} *MPC*, is greater than the achievable value by the propeller, n_{pr} *achievable*. Similarly, α *MPC*

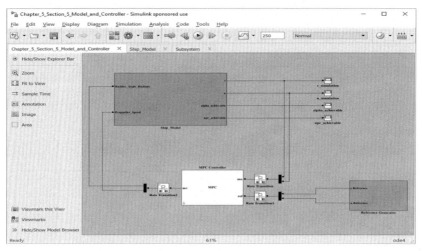

FIGURE 5.40 **Rate Transition** blocks.

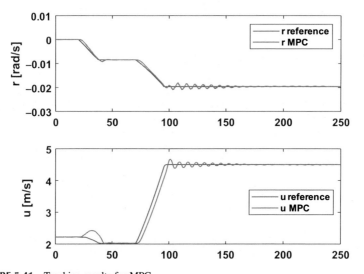

FIGURE 5.41 Tracking results for MPC.

is greater than α ***achievable***. This means that the soft constraints for the rate of the manipulated variables allowed MPC's solver to violate the maximum rate achievable by the actuators.

Fig. 5.41 shows oscillations in tracking the setpoints. In general, oscillations might be due to the external disturbance or due to the controller tuning. In this example, the oscillations are mainly due to the ocean's waves. Through tuning trials, the reader can find out whether the response of the

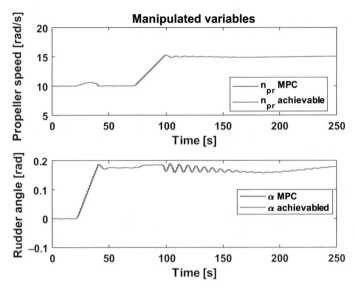

FIGURE 5.42   Manipulated variables for MPC.

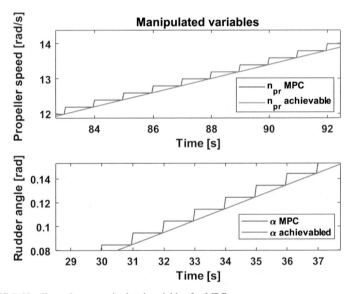

FIGURE 5.43   Zoom-in on manipulated variables for MPC.

system can be dampened further. Whether retuning is needed for MPC is dependent on the requirements. As a rule of thumb, pass the requirements without aggressively tuning the controller due to the robustness versus performance trade-off. Table 5.1 shows that the designed controller meets all the requirements.

TABLE 5.1 MPC Performance against the Requirements

Requirement	Simulation Value	Pass?
The maximum absolute error for the turning rate of the ship during the provided maneuver shall not exceed 4.5×10^{-3} rad/s	1.5×10^{-3}	Yes
The maximum absolute error for the ship's forward speed during the provided maneuver shall not exceed 0.42 m/s	0.3673	Yes
The root mean square error for the turning rate of the ship during the provided maneuver shall not exceed 2×10^{-3} rad/s	4.4252×10^{-3}	Yes
The root mean square error for the ship's forward speed during the provided maneuver shall not exceed 0.25 m/s	0.1167	Yes
The forward speed during the provided maneuver shall not exceed the reference forward speed by more than 0.5 m/s.	0.3055	Yes

To replicate the results of this section, the reader can run *Chapter_5_Section_7_Model_and_Controller.slx* which will simulate the final solution then run *Chapter_5_Section_7_Script.m* to plot results and display performance of the controller against requirements in the MATLAB *Command Window*.

5.8   APPLICATION PROBLEM: IMPACT OF TUNING ON ROBUSTNESS

In this application problem, the user will experience the impact of tuning on robustness of MPC. The same MPC controller that was developed in section 5.5 will be used. The performance of the controller will be assessed under harsher weather conditions. This means that the controller will face more unmeasured disturbance than the one used in the design. All the scripts, models and *.mat* files are provided in the material for *Chapter 5/Section_8*. The reader needs to download the material and set the working directory in MATLAB to be the location of the *Chapter 5/Section_8* folder.

1. Open *Chapter_5_Section_8_Model_and_Controller.slx* and run the simulation
2. Run *Chapter_5_Section_8_Script.m*
3. Did the controller pass all the requirements?
4. Retune the controller to improve the performance. To do so, open *mpcDesigner* and load *MPCDesignTask_Chapter_5_Section_8.mat* (hint: try changing the output weights). Once you are done with changing

the tuning, export the controller and save the *.mat* export under the name *MPC_Chapter_5_Section_8.mat* (type *save MPC_Chapter_5_Section_8 mpc1* in the command window).

REFERENCES

[1]   Automotive Control Systems, By A. Galip Ulsoy, Huei Peng, Melih Çakmakci Cambridge University Press, April 30, 2012.

[2]   http://www.theprojectmanagementblueprint.com/?p=109

Chapter 6

Multiple MPC Design for a Ship

6.1 INTRODUCTION

In many industrial applications, the dynamics of the plants are nonlinear. The questions that are asked frequently are: Is it highly nonlinear? Can we use a linear controller technique such as PID? In this chapter, the nonlinearity will be studied before the design of the controller. The response surface across the range of operation of the plant will be generated. This surface will give insight into the number of controllers needed for the full range.

Without loss of generality, the plant that is being considered is that of a ship. The process, however, applies to other plants. In Section 6.2, the operating range of the ship is defined. In Section 6.3, 20 simulations for the ship are carried out by using the *Parallel Computing Toolbox*. Methods such as changing Simulink parameters from the script, creating new file names and checking for MATLAB licenses are included in the script. In Section 6.4, the response surface is analyzed and nonlinearity is assessed. The linearized models of the ship for various operating ranges are created in Section 6.5. Using these models, multiple Model Predictive Control (MPC) controllers are created in Section 6.6. In Section 6.7, the multiple MPC controllers are integrated with the ship model. Major elements of the model are described and a switching logic for the controller is created. In Section 6.8, the model is simulated. Tracking results and controller performance is also analyzed. The problem of frequent switching of MPC modes is identified. A hysteresis block is introduced to mitigate the problem. The limitations of such an approach are highlighted. Finally, the chapter concludes with an application problem in Section 6.9 and references are listed in Section 6.10.

All the codes used in the chapter can be downloaded from *MATLAB File Exchange*. Follow the below link and search for the ISBN or title of the book.

https://www.mathworks.com/matlabcentral/fileexchange/

Alternatively, the reader can also download the material and other resources from the dedicated website or contact the authors for further help.

https://www.practicalmpc.com/

Practical Design and Application of Model Predictive Control.
DOI: https://doi.org/10.1016/B978-0-12-813918-9.00006-X

It is highly recommended to place the *Chapter_6* folder 5 levels under the root directory. Additionally, the reader shouldn't include any spaces in the names of the folders. Below is an example of where *Chapter_6* was placed four levels below the root directory without any spaces in the names of the folders:

C:\Users\Terminator\Documents\Chapter_6

6.2 DEFINING THE OPERATING REGIONS FOR THE SYSTEM

As we have seen in Chapter 4, System Identification for a Ship and Chapter 5, Single MPC Design for a Ship, the ship model has two inputs and two outputs of interest. The rudder angle α and propeller speed u are the inputs whereas the ship's turning rate and ship's speed are the outputs. The rudder angle can be moved from -0.3927 and 0.3927 radians respectively. Similarly, the propeller speed has a minimum speed of 5 and a 20 rad/s maximum speed. We will run a set of simulations that would cover the entire operating space. The rudder angle takes on the values of -0.3927, -0.1963, 0, 0.1963, and 0.3927 radians while the propeller speed, n_{pr}, takes on the values of 5, 10, 15, and 20 rad/s. This means that a total of 20 simulations will be run to generate the response surface. The steady state values of the ship forward speed, u, and turning rate, r, will be recorded. Fig. 6.1 shows the 20 points across the input space (rudder angle and propeller speed) where the response surface will be generated.

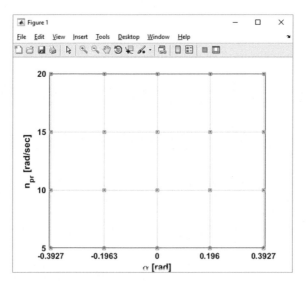

FIGURE 6.1   Input operating space of the ship.

6.3 STEADY STATE SIMULATIONS FOR THE OPERATING POINTS

If the reader has the *MATLAB Parallel Computing Toolbox*, simulations can be run in parallel to save run time. The scripts in the section can still be executed if the reader doesn't have the toolbox. The number of parallel simulations which can be initiated depends on the number of physical cores of the computer where *MATLAB* simulations are performed. The *Parallel Computing Toolbox* will assign the parallel jobs to several independent MATLAB instances, called workers. The Simulink model for the ship shown in Fig. 6.1 is used to run the actuator sweep simulations (same model for Chapter 4: System Identification for a Ship and Chapter 5: Single MPC Design for a Ship). Note that the inputs to the model are the two constant blocks representing α and n_{pr}. Their values in Fig. 6.2 are zeros, but the m script below will change their values. The outputs of the model that will be utilized to generate the response surface are the ship turning rate, *r_simulation*, and ship forward speed, *u_simulation*. The parameters *alpha_achievable* and *npr_achievable* are the position feedback for the rudder and propeller machines—the commanded value for these machines might differ from the actual value due to the physical constraints on the actuators. In this section, the commanded position and feedback are equal since we are providing values within the constraints of the actuators.

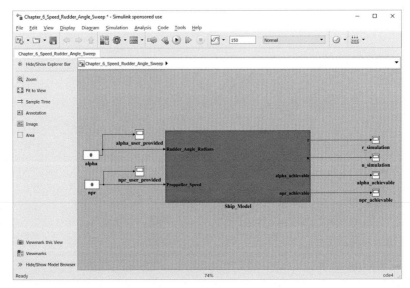

FIGURE 6.2   Ship Simulink model for actuator sweep.

For more details about the ship model, refer to Chapter 4, System Identification for a Ship.

The below m script, *Chapter_6_Section_3_1_Script.m*, checks if the user has the *Parallel Computing Toolbox* installed. There are two for loops—one to vary the propeller speed and the other to vary the rudder angle. The rudder angle is kept constant and the propeller speed is swept from 5 to 20 rad/s. In the next iteration, the rudder angle is changed and the propeller speed swept again from 5 to 20. Each job is given to the *MATLAB Parallel Computing Toolbox* using the *batch* command which will be distributed among MATLAB workers. If the Parallel Computing Toolbox is not installed, then the simulation run script is called in the normal *for* loop.

Chapter_6_Section_3_Script.m

```
%Book Title: Practical Design and Application of MPC
%Chapter: 6
%Section: 3
%Authors: Nassim Khaled and Bibin Pattel
%Last Modified: 6/14/2017
%%
% Distributed batch job simulation script
clc
clear all
bdclose all
% Check to see if the Parallel Computing Toolbox Exists
installed_toolboxes=ver;
[Toolbox_List{1:length(installed_toolboxes)}] = deal(installed_toolboxes.Name);
pct_toolbox_present = all(ismember('Parallel Computing Toolbox',Toolbox_List));
profile = 2;
% This loop creates the jobs
Job_Count = 1;
% Define the tables to speed for both Rudder angle and Propeller speed
Angles_to_Sweep = [-0.3927 -0.1963 0 0.1963 0.3927];
Speed_to_Sweep = [5 10 15 20];

% Create two for loops each for one actuator and set the loop iterations to
% the number of entries in the table. So we first fix the Rudder angle and
% vary the Propeller speed between 5-20.
for angle_index=1:5
for  speed_index =1:4
% Use Matlab's batch command to run the parallel simulations. Use
% batch only if Parallel Computing Toolbox is installeds
if(pct_toolbox_present)
            jobID(Job_Count) = ...
                batch(@Chapter_6_Section_3_2,... % Chapter_6_Section_3_2 is the
simulation run script
                1,... % Number of output arguments from the above simulation run script
                {Angles_to_Sweep(angle_index),Speed_to_Sweep(speed_index)},... % Cell
array of input arguments [ Selected Rudder angle and Propeller speed]
'Profile','local',... % Which parallel profile to use. local means the default profile.
'AutoAttachFiles',false,... % Turn off automatic attachment of files needed for
simulation
'AttachedFiles',{'Chapter_6_Section_3_2.m'},... % Attach the simulation run script for
the parallel workers
'CaptureDiary',true); % Capture teh worker Matlab's diary for debugging
            Job_Count = Job_Count+1; % Increment the job count
% If Parallel Computing Toolbox is not installed just use
% normal looping to run the simulations.
else

Chapter_6_Section_3_Function(Angles_to_Sweep(angle_index),Speed_to_Sweep(speed_index));
end
end
end
disp('Finished Submiting Job');
```

The m file which runs the simulation is implemented as a *function* which takes the specific rudder angle and propeller speed as arguments. The Simulink model is opened and the input argument values are set to the input constant block of the model shown in Fig. 6.2 using the *set_param* command. And then the simulation is run with the *sim* command and the results are saved to the desired results path.

Chapter_6_Section_3_Function.m

```
%Book Title: Practical Design and Application of MPC
%Chapter: 6
%Section: 3
%Authors: Nassim Khaled and Bibin Pattel
%Last Modified: 6/14/2017
%%
function return_val = Chapter_6_Section_3_Function(angle,speed)
% Define the Folder in which the simulation files are kept for each Matlab
% worker to access
sim_path = pwd;
addpath(genpath(sim_path));
% Define the Folder in which the simulation results will be stored at the
% end of the simulation by the workers
results_path = [pwd '\Results'];
if(~isdir(results_path))
    mkdir(results_path);
end
% Load the Model to be simulated
load_system('Chapter_6_Speed_Rudder_Angle_Sweep');
% Load the Base Workspace with Model Parameters
evalin('base','load data_model.mat');
% Set the steady state input actuator positions. set_param command is used
% to set the value of the constant blocks with what is passed in as
% argument to this script.
set_param('Chapter_6_Speed_Rudder_Angle_Sweep/alpha','Value',num2str(angle));
set_param('Chapter_6_Speed_Rudder_Angle_Sweep/npr','Value',num2str(speed));
% Set Sim options. Normally when a simulation is ran from a function using
% the sim command it runs in the function workspace. This opetion makes it
% run on the Matlab basse workspace
options = simset('SrcWorkspace','Base');
error_msg = '';
try
% Simulate the model and save the results into the results folder
    sim('Chapter_6_Speed_Rudder_Angle_Sweep',[],options);
    return_val = 1;
    fprintf(' Done.\n\n');
    save_str = ['save ' results_path '\Results_Par_Sim_Ship_Angle' num2str(angle)
'_Speed_' num2str(speed) '_Sweep.mat '];
    eval(save_str);
catch error_msg
% If there is any error occurred during the simulation catch the error
% message and write into a text file.
    return_val =0;
    error_msg.message
    fid = fopen([results_path '\Results_Par_Sim_Ship_Angle' num2str(angle) '_Speed_'
num2str(speed) '.txt'],'w+');
    fprintf(fid,'%s\n',error_msg.message);
    fclose(fid);
end
```

6.4 ANALYSIS OF STEADY STATE SIMULATIONS

If the simulations in Section 3 are executed properly, the *Results* folder should contain the .mat files for all the 20 rudder angle and propeller speed combinations as shown in Fig. 6.3.

The script, *Chapter_6_Section_4_Script.m*, loads the results that were generated in the previous section. The last value in the simulation is assumed to be the steady state value for the actuators and outputs. The script stores the data to arrays and plots them.

Chapter_6_Section_4_Function.m

```
%Book Title: Practical Design and Application of MPC
%Chapter: 6
%Section: 4
%Authors: Nassim Khaled and Bibin Pattel
%Last Modified: 10/15/2017
%%
clc
clear all
% Define the tables to speed for both Rudder angle and Propeller speed
Angles_to_Sweep = [-0.3927 -0.1963 0 0.1963 0.3927];
Speed_to_Sweep = [5 10 15 20];

curr_working folder = pwd;
% CD the folder Results.
cd Results
% Create two for loops each for one actuator and set the loop iterations to
% the number of entries in the table. So we first fix the Rudder angle and
% vary the Propeller speed between 5-20.
for angle_index=1:5
for  speed_index =1:4
% Load the results mat files one by one
        angle = Angles_to_Sweep(angle_index);
        speed = Speed_to_Sweep(speed_index);
        load_str = ['load ''Results\Results_Par_Sim_Ship_Angle' num2str(angle) '_Speed_'
num2str(speed) '_Sweep.mat '];
        eval(load_str);
%Grouping simulation data
        Measured_Outputs=[r_simulation(:,2) u_simulation(:,2)]; %Measured outputs
        Manipipulated_Variables=[alpha_achievable(:,2) npr_achievable(:,2)]; %Manipulated
variables
% Craete arrays for Inputs and Outputs. The last value of the 150
% Secs simulation is picked as the steady state point
        temp = Manipipulated_Variables(:,1);
        Rudder_Angle_Steady_State(angle_index,speed_index) = temp(end);
        temp = Manipipulated_Variables(:,2);
        Prop_Speed_Steady_State(angle_index,speed_index) = temp(end);
        temp = Measured_Outputs(:,1);
        Turning_Rate_Steady_State(angle_index,speed_index) = temp(end);
        temp = Measured_Outputs(:,2);
        Speed_Steady_State(angle_index,speed_index) = temp(end);

end
end
%% Plot the Actuator Sweep Results
figure;surf(Speed_to_Sweep,Angles_to_Sweep,Speed_Steady_State);
zlabel('u-measured [m/s]');
ylabel('\alpha [rad]');
xlabel('n_p_r [rad/s]');
set(gcf,'color',[1 1 1]);
grid on
figure;surf(Speed_to_Sweep,Angles_to_Sweep,Turning_Rate_Steady_State);
zlabel('r-measured [rad/s]');
ylabel('\alpha [rad]');
xlabel('n_p_r [rad/s]');
set(gcf,'color',[1 1 1]);
grid on
```

Figs. 6.4 and 6.5 show the ship forward speed and turning rate for various rudder angles and propeller speeds. From the surface plots, one can notice that the system is nonlinear across its operating region. More importantly, the system exhibits a change in sign of the controller gain. Referring to Fig. 6.4, it can be clearly seen that as the rudder angle α is increased from

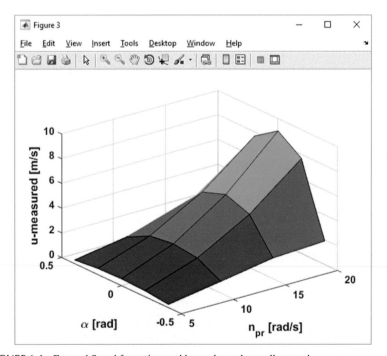

Name	Date modified	Type	Size
Results_Par_Sim_Ship_Angle0.1963_Speed_5_Sweep.mat	6/23/2017 3:26 PM	MATLAB Data	188 KB
Results_Par_Sim_Ship_Angle-0.1963_Speed_5_Sweep.mat	6/23/2017 3:25 PM	MATLAB Data	188 KB
Results_Par_Sim_Ship_Angle0.1963_Speed_10_Sweep.mat	6/23/2017 3:26 PM	MATLAB Data	183 KB
Results_Par_Sim_Ship_Angle-0.1963_Speed_10_Sweep.mat	6/23/2017 3:25 PM	MATLAB Data	183 KB
Results_Par_Sim_Ship_Angle0.1963_Speed_15_Sweep.mat	6/23/2017 3:26 PM	MATLAB Data	188 KB
Results_Par_Sim_Ship_Angle-0.1963_Speed_15_Sweep.mat	6/23/2017 3:25 PM	MATLAB Data	188 KB
Results_Par_Sim_Ship_Angle0.1963_Speed_20_Sweep.mat	6/23/2017 3:26 PM	MATLAB Data	195 KB
Results_Par_Sim_Ship_Angle-0.1963_Speed_20_Sweep.mat	6/23/2017 3:25 PM	MATLAB Data	195 KB
Results_Par_Sim_Ship_Angle0.3927_Speed_5_Sweep.mat	6/23/2017 3:27 PM	MATLAB Data	190 KB
Results_Par_Sim_Ship_Angle-0.3927_Speed_5_Sweep.mat	6/23/2017 3:24 PM	MATLAB Data	190 KB
Results_Par_Sim_Ship_Angle0.3927_Speed_10_Sweep.mat	6/23/2017 3:27 PM	MATLAB Data	187 KB
Results_Par_Sim_Ship_Angle-0.3927_Speed_10_Sweep.mat	6/23/2017 3:24 PM	MATLAB Data	187 KB
Results_Par_Sim_Ship_Angle0.3927_Speed_15_Sweep.mat	6/23/2017 3:27 PM	MATLAB Data	190 KB
Results_Par_Sim_Ship_Angle-0.3927_Speed_15_Sweep.mat	6/23/2017 3:24 PM	MATLAB Data	190 KB
Results_Par_Sim_Ship_Angle0.3927_Speed_20_Sweep.mat	6/23/2017 3:27 PM	MATLAB Data	193 KB
Results_Par_Sim_Ship_Angle-0.3927_Speed_20_Sweep.mat	6/23/2017 3:24 PM	MATLAB Data	193 KB
Results_Par_Sim_Ship_Angle0_Speed_5_Sweep.mat	6/23/2017 3:26 PM	MATLAB Data	189 KB
Results_Par_Sim_Ship_Angle0_Speed_10_Sweep.mat	6/23/2017 3:26 PM	MATLAB Data	184 KB
Results_Par_Sim_Ship_Angle0_Speed_15_Sweep.mat	6/23/2017 3:26 PM	MATLAB Data	188 KB
Results_Par_Sim_Ship_Angle0_Speed_20_Sweep.mat	6/23/2017 3:26 PM	MATLAB Data	192 KB

FIGURE 6.3 Steady state simulation results.

FIGURE 6.4 Forward Speed for various rudder angles and propeller speeds.

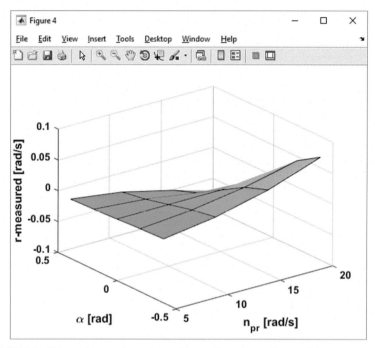

FIGURE 6.5 Ship turning rate for various rudder angles and propeller speeds.

-0.4 to 0 rad, the forward speed of the ship increases. But, as the rudder angle α increases from 0 to 0.4 rad, the forward speed of the ship decreases. A single mode MPC designed around one operating point will not be sufficient to control the system adequately in its entire operating region. We need to linearize the system around more operating points using the system identification technique learned in Chapter 5, Single MPC Design for a Ship. Using these linear models, we can design multiple MPCs.

6.5 CREATING LINEAR MODELS FOR THE ENTIRE OPERATING SPACE

There is not an exact formula to determine how many MPC models are needed for a plant. In the case of a ship, we know that there is a gain change for one of the actuators. As a minimum requirement, we would need one MPC controller for the range of the rudder that is negative and another for the positive. One can start with two MPC controllers and see whether the performance is adequate. The number can be increased until good performance is achieved. The main challenge that the designer will face is to

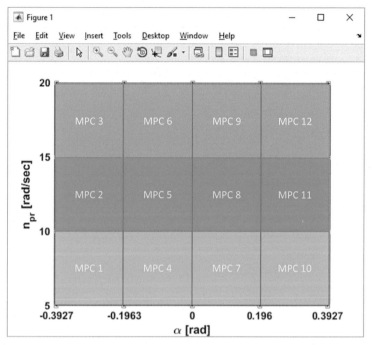

FIGURE 6.6   12 MPC controllers for the entire operational space.

decide whether performance is lacking due to a tuning problem or the insufficient number of MPCs.

To illustrate, we will use 12 MPC models. This allows us to demonstrate the automation process of designing any number of MPC controllers. Fig. 6.6 shows the 12 regions for which the MPC controllers will be developed. Note that the center of each region is where the nominal conditions for the controller will be assumed. For the step responses used for system identification, the actuators positions will start from the center of the region and be fired up to their maximum values in the region. Blocks of the same color denote regions at the same propeller speed.

The model of the ship will be used with steps input to run these simulations (Fig. 6.7). The rudder angle value will be stepped at time $t = 25$ seconds while the propeller speed value will be stepped at time $t = 75$ seconds for each operating point. The initial and final values for both actuators will be dictated by the values in Table 6.1. Since there are several simulations involved, we will make use of parallel computing.

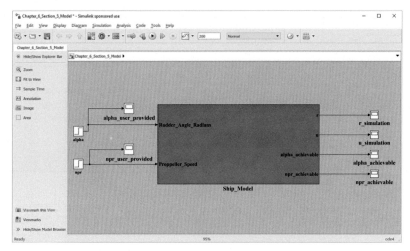

FIGURE 6.7 Ship model with step inputs.

TABLE 6.1 Actuator Step Values for System Identification

Operating Point #	Rudder Angle Initial Value (rad)	Rudder Angle Final Value (rad)	Propeller Speed Initial Value (rad/s)	Propeller Speed Final Value (rad/s)
1	− 0.2945	− 0.39265	7.5	12.5
2	− 0.2945	− 0.39265	12.5	17.5
3	− 0.2945	− 0.39265	17.5	22.5
4	− 0.09815	− 0.1963	7.5	12.5
5	− 0.09815	− 0.1963	12.5	17.5
6	− 0.09815	− 0.1963	17.5	22.5
7	0.09815	0.1963	7.5	12.5
8	0.09815	0.1963	12.5	17.5
9	0.09815	0.1963	17.5	22.5
10	0.2945	0.39265	7.5	12.5
11	0.2945	0.39265	12.5	17.5
12	0.2945	0.39265	17.5	22.5

The parallel simulation script, *Chapter_6_Section_5_1_Script*, executes the function *Chapter_6_Section_5_Function* through the *batch* command if *Parallel Computing Toolbox* is available, otherwise it directly calls *Chapter_6_Section_5_Function* with arguments of step initial and final values for the actuators.

Chapter_6_Section_5_1_Script.m

```
%Book Title: Practical Design and Application of MPC
%Chapter: 6
%Section: 5
%Authors: Nassim Khaled and Bibin Pattel
%Last Modified: 10/15/2017
%%
clc
clear all
% Check to see if the Parallel Computing Toolbox Exists
installed_toolboxes=ver;
[Toolbox_List{1:length(installed_toolboxes)}] = deal(installed_toolboxes.Name);
pct_toolbox_present = all(ismember('Parallel Computing Toolbox',Toolbox_List));
profile = 2;
% This loop creates the jobs
Job_Count = 1;
% Define the initial and final step values for the Rudder angle input
Angle_Step_Initial_Values = [-0.2945    -0.09815    0.09815    0.2945];
Angle_Step_Final_Values =    [-0.39265   -0.1963     0.1963     0.39265];
% Define the initial and final step values for the Propeller speed input
Speed_Step_Initial_Values = [7.5 12.5 17.5 ];
Speed_Step_Final_Values = [12.5 17.5 22.5 ];
% Create two for loops each one for one actuator and set the loop iterations to
% the number of entries in the table. So we first step the Rudder angle at time t = 25
Secs and
% then step Propeller speed at time t = 75 Secs.
for angle_index=1:4
for  speed_index =1:3
% Use Matlab's batch command to run the parallel simulations. Use
% batch only if Parallel Computing Toolbox is installeds
if(pct_toolbox_present)
            jobID(Job_Count) = ...
                batch(@Chapter_6_Section_5_Function,... % Chapter_6_Section_5_Function is
the simulation run script
                1,... % Number of output arguments from the above simulation run script

{Angle_Step_Initial_Values(angle_index),Angle_Step_Final_Values(angle_index), ...

Speed_Step_Initial_Values(speed_index),Speed_Step_Final_Values(speed_index)},... % Cell
array of input arguments [ Selected Rudder angle and Propeller speed]
'Profile','local',... % Which parallel profile to use. local means the default profile.
'AutoAttachFiles',false,... % Turn off automatic attachment of files needed for
simulation
'AttachedFiles',{'Chapter_6_Section_5_Function.m'},... % Attach the simulation run script
for the parallel workers
'CaptureDiary',true); % Capture teh worker Matlab's diary for debugging
            Job_Count = Job_Count+1; % Increment the job count
% If Parallel Computing Toolbox is not installed just use
% normal looping to run the simulations.
else

Chapter_6_Section_5_Function(Angle_Step_Initial_Values(angle_index),Angle_Step_Final_Valu
es(angle_index), ...

Speed_Step_Initial_Values(speed_index),Speed_Step_Final_Values(speed_index));
end
end
end
disp('Finished Submiting Job');
```

Chapter_6_Section_5_Function is the function that runs the Simulink model with actuators step inputs and saves the results.

Chapter_6_Section_5_Function.m

```
%Book Title: Practical Design and Application of MPC
%Chapter: 6
%Section: 5
%Authors: Nassim Khaled and Bibin Pattel
%Last Modified: 10/15/2017
%%
function return_val =
Chapter_6_Section_5_Function(angle_step_initial,angle_step_final,speed_step_initial,speed
_step_final)
% Define the Folder in which the simulation files are kept for each Matlab
% worker to access
sim_path = pwd;
addpath(genpath(sim_path));
% Define the Folder in which the simulation results will be stored at the
% end of the simulation by the workers
results_path = [sim_path '\Results_Multi_Model_Sys_ID'];
if(~isdir(results_path))
    mkdir(results_path);
end
% Load the Model to be simulated
load_system('Chapter_6_Section_5_Model');
% Load the Base Workspace with Model Parameters
evalin('base','load data_model.mat');
% Set the steady state input actuator positions. set_param command is used
% to set the value of the constant blocks with what is passed in as
% argument to this script.
set_param('Chapter_6_Section_5_Model/alpha','Before',num2str(angle_step_initial),'After',
num2str(angle_step_final));
set_param('Chapter_6_Section_5_Model/npr','Before',num2str(speed_step_initial),'After',nu
m2str(speed_step_final));
% Set Sim options. Normally when a simulation is ran from a function using
% the sim command it runs in the function workspace. This option makes it
% run on the Matlab base workspace
options = simset('SrcWorkspace','Base');
error_msg = '';
try
% Simulate the model and save the results into the results folder
    sim('Chapter_6_Section_5_Model',[],options);
    return_val = 1;
    fprintf(' Done.\n\n');
    save_str = ['save ' results_path '\Results_Par_Sim_Ship_Angle_Step_'
num2str(angle_step_initial) '_' num2str(angle_step_final) '_Speed_Step_'
num2str(speed_step_initial) '_' num2str(speed_step_final) '.mat '];
    eval(save_str);
catch error_msg
% If there is any error occured during the simulation catch the error
% message and write into a text file.
    return_val =0;
    error_msg.message
    fid = fopen([results_path '\Results_Par_Sim_Ship_Angle_Step'
num2str(angle_step_initial) '_' num2str(angle_step_final) '_Speed_Step_'
num2str(speed_step_initial) '_' num2str(speed_step_final) '.txt'],'w+');
    fprintf(fid,'%s\n',error_msg.message);
    fclose(fid);
end
```

To obtain the system identification data, run the *Chapter_6_Section_5_1_Script.m*. Once it finishes execution, the results will be saved in the folder *Results_Multi_Model_Sys_ID* (Fig. 6.8).

After finishing the simulations and collecting the step response data, the system identification process discussed in Chapter 5, Single MPC Design for a Ship, can be executed for the 12 simulation results to identify the linear models.

Chapter_6_Section_5_2_Script.m loads the simulation results, prepares the data, runs *n4sid* and *pem* commands for identification, and saves the linear models in to a *.mat* file and plots how good the fit is in form of percentage.

Chapter_6_Section_5_2_Script.m

```
%Book Title: Practical Design and Application of MPC
%Chapter: 6
%Section: 5
%Authors: Nassim Khaled and Bibin Pattel
%Last Modified: 10/15/2017
%%
clc
clear all
plot_enable =1;
Mode_Num_Sys_ID =1;
% Define the initial and final step values for the Rudder angle input
Angle_Step_Initial_Values = [-0.2945    -0.09815    0.09815    0.2945];
Angle_Step_Final_Values = [-0.39265    -0.1963    0.1963    0.39265];
% Define the initial and final step values for the Propeller speed input
Speed_Step_Initial_Values = [7.5  12.5 17.5 ];
Speed_Step_Final_Values =    [12.5 17.5 22.5 ];
% Specify the path to look for the simulation results
results_path = [pwd '\Results_Multi_Model_Sys_ID'];
% Create two for loops each for one actuator and set the loop iterations to
% the number of entries in the table. So we first step the Rudder angle at time t = 25
Secs and
% then step Propeller speed at time t = 75 Secs.
for angle_index_new=1:4
for  speed_index_new =1:3
        angle_step_initial = Angle_Step_Initial_Values(angle_index_new);
        angle_step_final = Angle_Step_Final_Values(angle_index_new);
        speed_step_initial = Speed_Step_Initial_Values(speed_index_new);
        speed_step_final = Speed_Step_Final_Values(speed_index_new);
        load_str = ['load ' results_path '\Results_Par_Sim_Ship_Angle_Step_'
num2str(angle_step_initial) '_' num2str(angle_step_final) '_Speed_Step_'
num2str(speed_step_initial) '_' num2str(speed_step_final) '.mat '];
        eval(load_str);
        clear sys1sys2
        %% Plotting Settings
        set(0,'DefaultLineLineWidth',1)
        set(0,'DefaultAxesFontWeight','bold')
        set(0,'DefaultAxesFontName','Arial')
        set(0,'DefaultAxesFontSize',12)
        %% Plot the propeller speed
        stp_sz = 0.0500;
        plot_enable =1;
if(plot_enable)
        figure
% Plot the rudder angle
        subplot(411)
        plot(alpha_user_provided (:,1),alpha_user_provided(:,2))
        hold on
        plot(alpha_achievable(:,1),alpha_achievable(:,2))
        legend('\alpha command [rad]','\alpha measured [rad]')
        grid on
% Plot the propeller speed
        subplot(412)
        plot(npr_user_provided(:,1),npr_user_provided(:,2))
        hold on
        plot(npr_achievable(:,1),npr_achievable(:,2))
        legend('n_p_r command [rad/s]','n_p_r measured [rad/s]')
        grid on
% Plot the ship heading
        subplot(413)
        plot(r_simulation(:,1),r_simulation(:,2))
        legend('r [rad/s]')
        grid on
% Plot the ship speed
        subplot(414)
        plot(u_simulation(:,1),u_simulation(:,2))
        legend('u m/s')
        xlabel('Time [s]')
        grid on

% Get the nominal values for inputs and outputs. This will be
% required later for the controller design. Taking the 300'th
% data sample because thats when the output may have reached
% steady state when sim time is 15 Secs .(300 * 0.05 = 15 Secs)
        temp = alpha_achievable(:,2);
```

```
        Inputs(1).Nominal = temp(300);
        temp = npr_achievable(:,2);
        Inputs(2).Nominal = temp(300);

        temp = r_simulation(:,2);
        Outputs(1).Nominal = temp(300);
        temp = u_simulation(:,2);
        Outputs(2).Nominal = temp(300);


end
        %% Grouping simulation data and neglecting first few seconds(Nstart*stp_sz)
        Nstart = 100; %Clip the data starting from Nstart as the first few seconds of the
simulation should be discarded

        Measured_Outputs=[r_simulation(Nstart:end,2) u_simulation(Nstart:end,2)];
%Measured outputs
        Manipipulated_Variables=[alpha_achievable(Nstart:end,2)
npr_achievable(Nstart:end,2)]; %Manipulated variables

        %% Obtaining initial conditions at step time=Nstart
        Measured_Outputs_Nstart=Measured_Outputs(Nstart,:); %Capturing the measured
outputs at step time= Nstart
        Manipipulated_Variables_Nstart=Manipipulated_Variables(Nstart,:); %Capturing the
manipulated variables at step time= Nstart

        %% Forcing response to start from zero initial conditions
        Measured_Outputs_zero_initial_conditions=Measured_Outputs-
repmat(Measured_Outputs_Nstart,length(Measured_Outputs),1); %Subtracting initial
conditions for measured outputs to obtain zero response at step time= Nstart
        Manipipulated_Variables_zero_initial_conditions=Manipipulated_Variables-
repmat(Manipipulated_Variables_Nstart,length(Manipipulated_Variables),1); %Subtracting
initial conditions for
manipulated variables to obtain zero actuation at step time= Nstart

        %% Prepare date for system identification

data=iddata(Measured_Outputs_zero_initial_conditions,Manipipulated_Variables_zero_initial
_conditions,stp_sz); %data is packaged for system identification using iddata

        %% Generate a preliminary 4th order system that fits the data
        sys1=n4sid(data,3,'Form','canonical','DisturbanceModel','None','InputDelay',[0
0]','InitialState','zero'); %n4sid generates a preliminary system in the canonical form
%with zero disturbance, zero delay and zero initial conditions

        %% Generate a more refined system
        sys2 = pem(data,sys1,'InitialState','zero') %pem generates a more refined system
that fits the data better

        %% Define the options for comparing the various identified systems
        opt = compareOptions('InitialCondition','Z');
        [Y,fit,x0]=compare(data,sys2);
        Y_1=Y.OutputData(:,1);
        Y_2=Y.OutputData(:,2);

        %% Plotting the system identification results
if(plot_enable)
        figure
% figure('Position',[358 0 500 1100])
        subplot(4,1,1)
        plot(0:stp_sz:stp_sz*(length(Y_1)-
1),Manipipulated_Variables_zero_initial_conditions(:,1))
        grid on
        legend('\alpha [rad]')
        subplot(4,1,2)
        plot(0:stp_sz:stp_sz*(length(Y_1)-
1),Manipipulated_Variables_zero_initial_conditions(:,2))
        grid on
        legend('n_p_r [rad/s]')
        subplot(4,1,3)
        plot(0:stp_sz:stp_sz*(length(Y_1)-
1),Measured_Outputs_zero_initial_conditions(:,1))
        grid on
        hold on
        plot(0:stp_sz:stp_sz*(length(Y_1)-1),Y_1,'r')
        legend('r measured [rad/s]','r system identification [rad/s]')
```

```
            subplot(4,1,4)
            plot(0:stp_sz:stp_sz*(length(Y_1)-
1),Measured_Outputs_zero_initial_conditions(:,2))
            grid on
            hold on
            plot(0:stp_sz:stp_sz*(length(Y_1)-1),Y_2,'r')
            legend('u measured [m/s]','u system identification [m/s]')
            xlabel('Time [s]')
end
        %% Compute the goodness of fit [%]

gof_Sys_ID_r(Mode_Num_Sys_ID)=goodnessOfFit(Measured_Outputs_zero_initial_conditions(:,2)
,Y_2,'NRMSE')*100 % gof for r

gof_Sys_ID_u(Mode_Num_Sys_ID)=goodnessOfFit(Measured_Outputs_zero_initial_conditions(:,1)
,Y_1,'NRMSE')*100 % gof for u

        %% Assess stability of the identified system
        pole(sys2) %poles of sys2
% Resample the discrete model to 1 Sec because the controller will
% be using the 1 Sec sampled model
        sys2 = d2d(sys2,1);
        eval_str = ['Ship_Linear_Model_Mode_' num2str(Mode_Num_Sys_ID) ' = sys2;'];
        eval(eval_str);
        eval_str = ['save Ship_Linear_Model_Mode_' num2str(Mode_Num_Sys_ID) '.mat''
Ship_Linear_Model_Mode_' num2str(Mode_Num_Sys_ID) ' Inputs Outputs '];
        eval(eval_str);
        Mode_Num_Sys_ID = Mode_Num_Sys_ID+1;
end
end

%% Plot the goodness of fit for the 12 models
figure;
set(gcf,'color',[1 1 1 ]);
subplot(211);
plot(gof_Sys_ID_r,'linewidth',2,'Marker','Diamond');
hold on
grid on
plot(75+0.*gof_Sys_ID_r,'r-.');
xlabel('Modal Point')
ylabel('Goodness of Fit [%]')
title('Ship Turning Rate')
ylim([0 100])
xlim([1 12])

grid on
subplot(212);
plot(gof_Sys_ID_u,'linewidth',2,'Marker','Diamond');
hold on
grid on
plot(75+0.*gof_Sys_ID_u,'r-.');
xlabel('Modal Point')
ylabel('Goodness of Fit [%]')
title('Ship Forward Speed')
ylim([0 100])
xlim([1 12])
```

Fig. 6.9 shows the system identification data for mode 1. The top plot shows rudder angle command and measurement. The rudder machine cannot move instantaneously to match the command and thus the difference in the command and measured position of the rudder. Similarly, the second plot

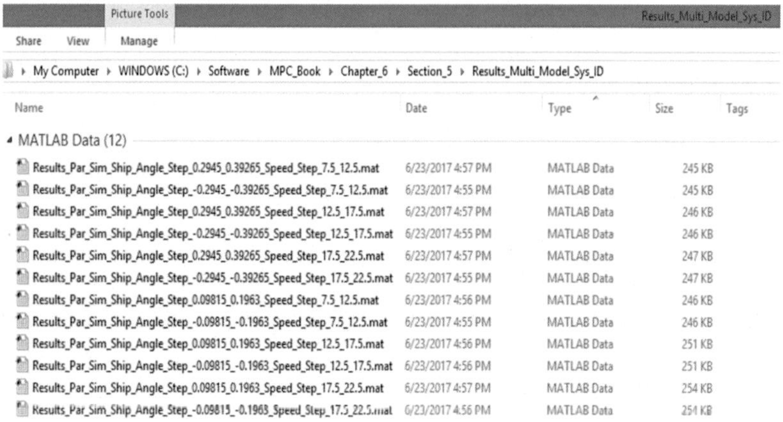

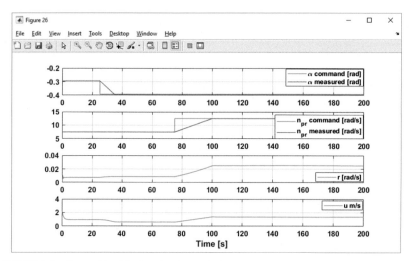

FIGURE 6.8 Step response results.

FIGURE 6.9 System identification data for mode 1.

from top of Fig. 6.9 shows the propeller command and feedback. The third plot from top shows the turning rate response. The turning rate responds to both rudder step and propeller step. The bottom plot of Fig. 6.9 shows the forward speed of the ship. It also responds to both actuators.

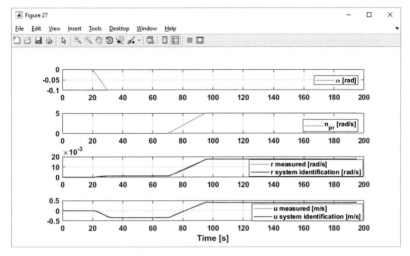

FIGURE 6.10 System identification results for mode 1.

Fig. 6.10 shows the results of the system identification. The total time of Fig. 6.10 is shorter by 5 seconds compared with Fig. 6.9 as we cropped the transient portion of the initial phase of the simulation. The first few seconds of the simulations are discarded as the simulation model might not have the proper initial conditions and needs a few seconds to stabilize.

Comparing the rudder angle at time zero for Figs. 6.9 and 6.10, the reader can notice that the rudder angle in 6.10 starts from zero. Similarly, the propeller speed, turning rate, and forward speed all start from zero in Fig. 6.10. We have offset the inputs and outputs so that they start from zero. These offset values for the actuators and outputs will be used in the design of the MPC controller as the nominal values.

We use *gof* metric to quantify the quality of the fit. It is described in Eq. (4.1) and it relies on the *goodnessOfFit* function in *MATLAB* with *NRMSE* (Normalized Root Mean Square Error). Results of the fit are shown in Fig. 6.11. All the *gof* are above the recommended 75% threshold that we proposed in Chapter 5, Single MPC Design for a Ship, and are highlighted in red.

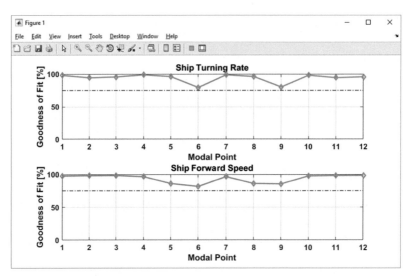

FIGURE 6.11 Goodness of fit for the 12 models.

6.6 DESIGNING A MULTIMODE MPC

Using the 12 linear models from the previous section, we are going to design 12 MPC controllers. These controllers will be used as a multimode switched MPC. In Chapter 5, Single MPC Design for a Ship, we detailed the process of designing a single mode MPC using the MPC designer GUI. Since multiple MPC controllers need to be designed in this section, we will choose MATLAB scripting to design the controllers. This is highly recommended especially when trying to design multiple MPC's or when automating the design process.

Chapter_6_Section_6_Script loads linear models one by one in a loop, designs a MPC controller for that mode, and saves the controller object as a *MATLAB .mat* file in the working folder. The name of the saved *.mat* file and the name of the MPC object inside the mat file will be the same (e.g., *MPC_Controller_Mode_1*). Fig. 6.12 shows the resulting files. All the 12 MPC designs passed the *Design Review* tests. Fig. 6.13 shows the resulting *MATLAB* webpages.

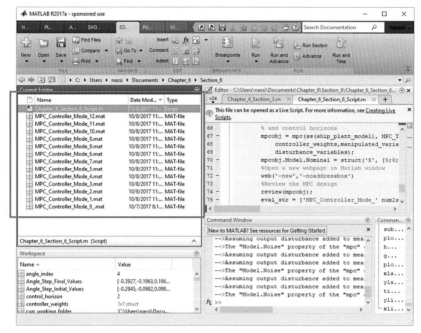

FIGURE 6.12   Output files for the MPC design script.

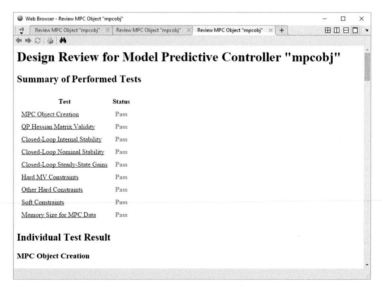

FIGURE 6.13   Design review results for the MPC controller.

Chapter_6_Section_6_Script.m

```
%Book Title: Practical Design and Application of MPC
%Chapter: 6
%Section: 6
%Authors: Nassim Khaled and Bibin Pattel
%Last Modified: 10/15/2017
%%
clc
clear all
% Define the initial and final step values for the Rudder angle input
Angle_Step_Initial_Values = [-0.2945    -0.09815    0.09815    0.2945];
Angle_Step_Final_Values =   [-0.39265   -0.1963     0.1963     0.39265];
% Define the initial and final step values for the Propeller speed input
Speed_Step_Initial_Values = [7.5  12.5 17.5 ];
Speed_Step_Final_Values =   [12.5 17.5 22.5 ];
% Specify the path to look for the simulation results
curr_working_folder = pwd;
cd('..');
linear_model_results_path = [pwd '\Section_5'];
cd(curr_working_folder);


Mode_Num_Sys_ID =1;
% MPC Controller sampling time
MPC_Ts = 1;
% Define the Prediction and Control Horizons
prediction_horizon = 10;
control_horizon = 2;
Mode_Num_Sys_ID =1;
% Define the MPC Controller Weights
controller_weights = struct('ManipulatedVariables',[0 0],...
'ManipulatedVariablesRate',[0.30 0.30],...
'OutputVariables',[0.30 0.30],...
'ECR',3.3201e+04);
% Define the MPC Controller Manipulated variable settings
manipulated_variables = struct('Min',{-0.3900 -30},...
'Max',{0.3900 30},...
'MinECR',{0 0},...
'MaxECR',{0 0},...
'RateMin',{-Inf -Inf},...
'RateMax',{Inf Inf},...
'RateMinECR',{0.0100 0.0100},...
'RateMaxECR',{0.1000 0.1000},...
'Target',{'nominal', 'nominal'},...
'Name',{'u1','u2'},...
'Units',{'rad','rad/sec'},...
'ScaleFactor',{0.7800 25});
% Define the MPC Controller Output variable settings
output_variables = struct('Min',{-3 -20},...
'Max',{3 20},...
'MinECR',{1 1},...
'MaxECR',{1 1},...
'Name',{'y1','y2'},...
'Units',{'rad/s','m/sec'},...
'ScaleFactor',{0.0400 25});
% No Disturbance Model
disturbance_variables = [];
% Create two loops for two actuator data
for angle_index=1:4
for speed_index =1:3
    % Load the system id linear ship models created in Section 6.5
    load_str = ['load ' linear_model_results_path '\Ship_Linear_Model_Mode_'
    num2str(Mode_Num_Sys_ID) '.mat'' Ship_Linear_Model_Mode_' num2str(Mode_Num_Sys_ID) '
    Inputs Outputs '];
    eval(load_str);
    % Get the plant model
    ship_plant_model = eval(['Ship_Linear_Model_Mode_' num2str(Mode_Num_Sys_ID)]);
    ship_plant_model = setmpcsignals(ship_plant_model,'MV',[1 2],'UD',[]);
    % Create the MPC controller for the plant with default prediction and control horizons
    mpcobj = mpc(ss(ship_plant_model), MPC_Ts,prediction_horizon,control_horizon,...
    controller_weights,manipulated_variables,output_variables,...
    disturbance_variables);
    mpcobj.Model.Nominal = struct('X', [0;0;0], 'U',
    [Inputs(1).Nominal;Inputs(2).Nominal], 'Y', [Outputs(1).Nominal;Outputs(2).Nominal],
```

```
'DX', [0 ;0; 0]);
%Open a new webpage in Matlab window
web('-new','-noaddressbox')
%Review the MPC design
review(mpcobj);
eval_str = ['MPC_Controller_Mode_' num2str(Mode_Num_Sys_ID) ' = mpcobj;'];
eval(eval_str);
eval_str = ['save MPC_Controller_Mode_' num2str(Mode_Num_Sys_ID) '.mat''
MPC_Controller_Mode_' num2str(Mode_Num_Sys_ID) ];
eval(eval_str);
Mode_Num_Sys_ID= Mode_Num_Sys_ID + 1;
end
end
```

6.7   SIMULINK MODEL FOR MULTIPLE MPC

In this section, we will describe the Simulink model for a closed loop system of a ship with multiple MPC controllers. Main blocks will be described briefly as well as how to get them from the Simulink library. It is advised that the reader opens *Chapter_6_Section_7_Model_1.slx* from *Chapter_6/Section_7* folder while reading through this section. The main components of the model are:

- Reference generation system
- Ship plant system
- Ship controller system
- Reference tracking and logging

A top-level view of the simulation model is shown in Fig. 6.14.

The reference generation system provides the controller with a reference signal for both the ship's turning rate and the forward speed. This maneuver is designed to push the controller across multiple modes. Fig. 6.15 shows a time series plot for the maneuver. Figs. 6.16 and 6.17 show the Simulink

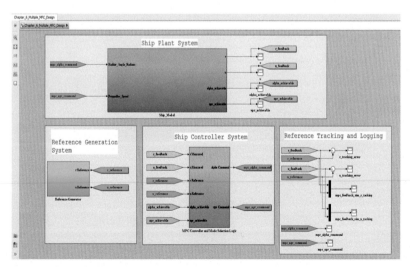

FIGURE 6.14   Top-level Simulink model of a multiple MPC controller for the ship.

subsystem for the reference generation. For readers interested in a guidance system for ships, check the United States patent by Chalhoub and Khaled [1] or the related publication [2].

Fig. 6.18 shows the top-level view for a ship MPC controller model. It takes the measured outputs *r_feedback*, *u_feedback* from the ship's plant

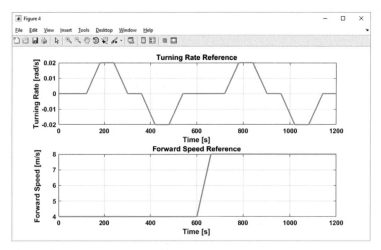

FIGURE 6.15   Reference signals for the MPC controller.

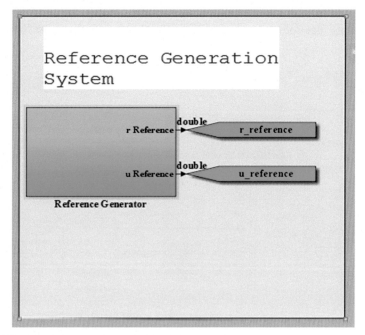

FIGURE 6.16   Reference generator top-level view.

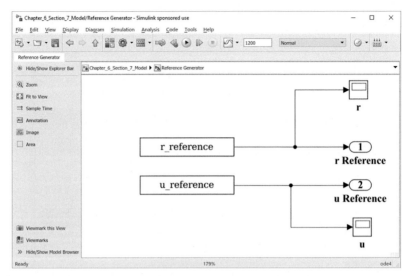

FIGURE 6.17 Reference generator subsystem.

subsystem, and controller references *r_reference* and *u_reference* from the reference generation subsystem. It also takes *alpha_achievable*, *npr_achievable* which is the real-time feedback of the current actuator positions. The actuator positions will be used to select which controller will be used among the 12 MPC controllers. The selection will be made based on where the actuators positions fall in Fig. 6.6. The integer representing the region for the selected MPC controller is called mode.

The outputs of the controller block are the rudder angle and propeller speed commands. Inside the controller block of Fig. 6.18 there are two main parts: Mode selection and the MPC controllers. Fig. 6.19 shows these two main blocks, *MPC_Mode_Selection* and *Multiple_MPC_Controllers*. Notice the rate transition blocks for the inputs as well as the outputs in this block. These are added to the controller since the measurements are updated at 0.05 seconds (step size of the simulation) while the MPC controller executes at 1 second (refer to Chapter 5: Single MPC Design for a Ship for details).

MPC_Mode_Selection is implemented using a MATLAB function block that can be obtained from *Simulink/User-Defined Functions/MATLAB Function*. Users can double click on the block to open the m file editor and type in the logic (Figs. 6.20 and 6.21).

The Simulink MPC library can be opened by typing *mpclib* on the MATLAB command window. Alternatively, it can be found in the Simulink library by searching for MPC. The MPC library shown in Fig. 6.22 and the *Multiple MPC* block is highlighted. Note that we unchecked the measured disturbance (md) option in the Multiple MPC block as we aren't using any

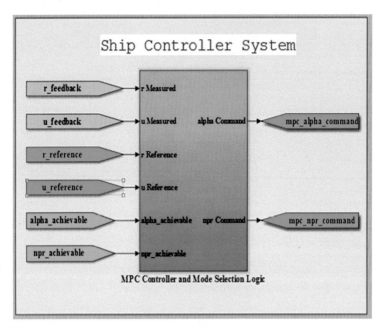

FIGURE 6.18   Ship controller subsystem top view.

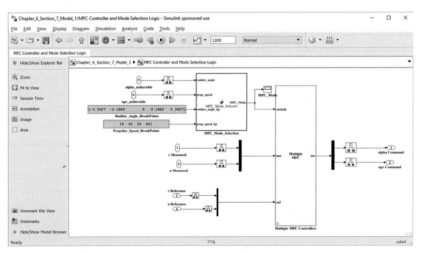

FIGURE 6.19   Ship controller subsystem inside view.

md in this controller simulation. Fig. 6.23 shows the cell array of the MPC Controllers field and MPC Initial Controller States. As shown in Fig. 6.18, the multiple MPC block has now three inputs and one output. The first input is the MPC *switch* value from the mode selection logic, the second

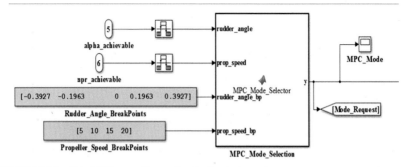

FIGURE 6.20 Ship MPC controller mode selection logic.

FIGURE 6.21 Function block for MPC selector logic.

input is the *measured output (mo)* which is a *muxed* signal from the ship plant, and the third input is the *reference (ref)* which is the controller references *muxed* together. The output of the multiple MPC controller block is the *manipulated variables (mv)* which will be *demuxed* and fed back to the ship actuator input.

The ship plant model is the same one we used in Section 6.5 for linearization and system identification (Fig. 6.24).

Fig. 6.25 shows the logging and scoping of tracking errors. Additionally, the MPC command for the actuators references is logged and scoped. There are other signals logged to the workspace through the scopes

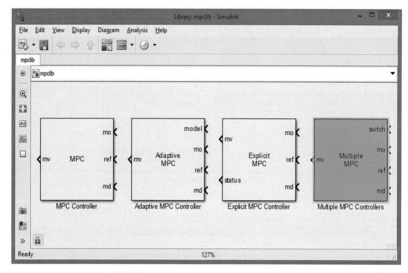

FIGURE 6.22   MPC Simulink library.

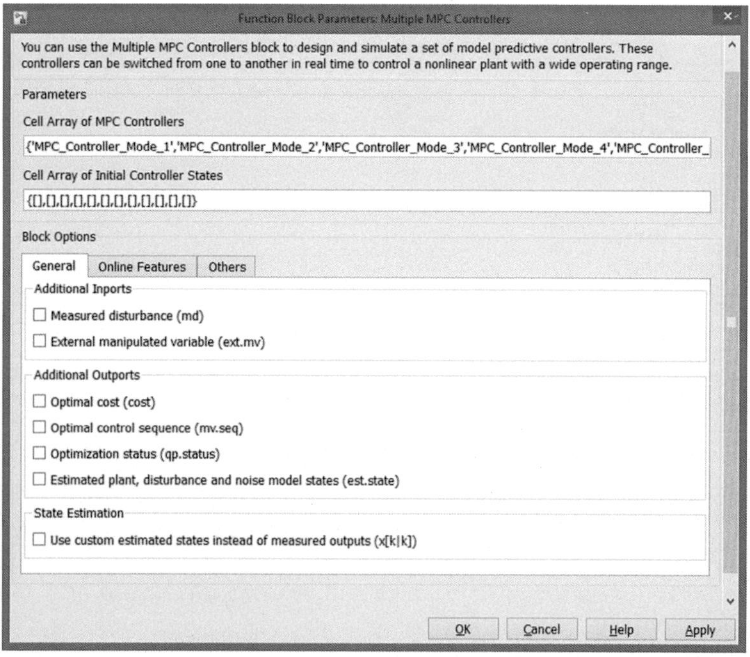

FIGURE 6.23   Settings for multiple MPC block.

in the Simulink model, but are not used in this section. The ship's turning rate and forward speed in addition to the actual feedback for the position of the actuators are also logged to the workspace (through the scopes in Fig. 6.24).

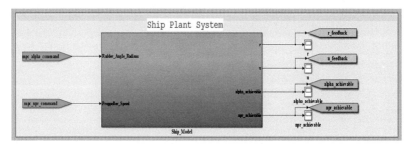

FIGURE 6.24   Ship plant subsystem for multiple MPC controllers.

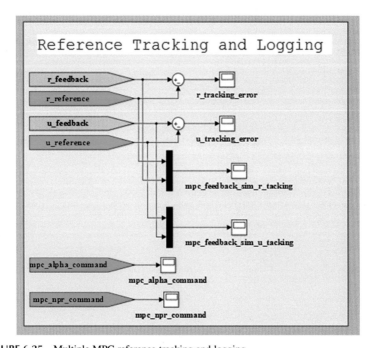

FIGURE 6.25   Multiple MPC reference tracking and logging.

6.8   MULTIPLE MPC CONTROLLER SIMULATION RESULTS

To simulate the model discussed in Section 6.7, *Chapter_6_Section_7_Model_1.slx*, download *Chapter_6/Section_8* material. Change the MATLAB directory to the downloaded folder. Execute the script *Chapter_6_Section_8_Script_1.m*. The script will load the plant parameters for the ship in addition to the MPC controllers. It will simulate *Chapter_6_Section_7_Model_1.slx* and plot the results.

Chapter_6_Section_8_Script_1.m

```
%Book Title: Practical Design and Application of MPC
%Chapter: 6
%Section: 8
%Authors: Nassim Khaled and Bibin Pattel
%Last Modified: 10/15/2017
%%
warning off
clc
clear all
bdclose all
% Open the Model
load_system('Chapter_6_Section_7_Model_1');
% Load the Base Workspace with Model Parameters
load data_model.mat
% Initialize the MPC Controller Objects
load('MPC_Controller_Mode_12.mat')
load('MPC_Controller_Mode_11.mat')
load('MPC_Controller_Mode_10.mat')
load('MPC_Controller_Mode_9.mat')
load('MPC_Controller_Mode_8.mat')
load('MPC_Controller_Mode_7.mat')
load('MPC_Controller_Mode_6.mat')
load('MPC_Controller_Mode_5.mat')
load('MPC_Controller_Mode_4.mat')
load('MPC_Controller_Mode_3.mat')
load('MPC_Controller_Mode_2.mat')
load('MPC_Controller_Mode_1.mat')
% Load the MPC Reference for the Controller to Track
load Test_SS_Target_V1.mat
% Simulate the Model
sim('Chapter_6_Section_7_Model');
%% Plot the Results
close all
figure
set(gcf,'color',[1 1 1]);
subplot(211);
plot(mpc_feedback_sim_r_tacking(:,1),mpc_feedback_sim_r_tacking(:,2),'linewidth',2);
hold all
plot(mpc_feedback_sim_r_tacking(:,1),mpc_feedback_sim_r_tacking(:,3),'linewidth',2);
grid on
legend('r Reference','r Measured');
xlabel('Time [s]');
ylabel('[rad/s]');
title('Turning Rate Tracking');

subplot(212);
plot(mpc_feedback_sim_u_tacking(:,1),mpc_feedback_sim_u_tacking(:,2),'linewidth',2)
hold all
plot(mpc_feedback_sim_u_tacking(:,1),mpc_feedback_sim_u_tacking(:,3),'linewidth',2)
grid on
legend('u Reference','u Measured');
xlabel('Time Secs');
ylabel('[m/s]');
title('Forward Speed Tracking');
%%
figure
set(gcf,'color',[1 1 1]);
subplot(211);
plot(mpc_alpha_command(:,1),mpc_alpha_command(:,2),'linewidth',2);
hold on
plot(alpha_achievable(:,1),alpha_achievable(:,2),'r','linewidth',2);
legend('\alpha MPC Command','\alpha Measured')

grid on
xlabel('Time [s]');
ylabel('[rad/s]');
title('Rudder Angle');
```

```
subplot(212);
plot(mpc_npr_command(:,1),mpc_npr_command(:,2),'linewidth',2);
hold on
plot(npr_achievable(:,1),npr_achievable(:,2),'r','linewidth',2);
legend('n_p_r MPC Command','n_p_r Measured')
grid on
xlabel('Time [s]');
ylabel('[rad/s]');
title('Propeller Speed');

%%
figure
set(gcf,'color',[1 1 1]);
plot(MPC_Mode(:,1),MPC_Mode(:,2),'linewidth',2);
grid on
xlabel('Time Secs');
title('Selected MPC Mode');
```

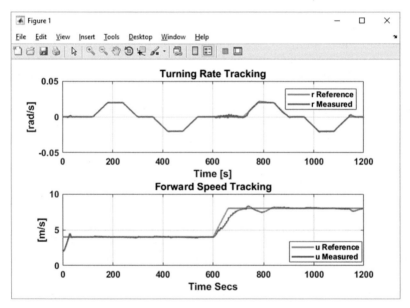

FIGURE 6.26 Multiple MPC controller tracking.

Fig. 6.26 shows the MPC controller tracking for the ship's turning rate and forward speed. Beside the chattering at about 650 seconds into the simulation, the turning rate tracking is good. Forward speed tracking is sluggish especially from 600 to 800 seconds. Fig. 6.27 gives more insight into the reason for the chattering. In specific terms, at the beginning of the simulation and from 600 to 700 seconds, the rudder command from the MPC exhibits a chattering behavior. Similarly, the propeller command exhibits the same behavior from 600 to 700 seconds. One main potential cause for such a performance in a multiple MPC is the logic that the commands are switching

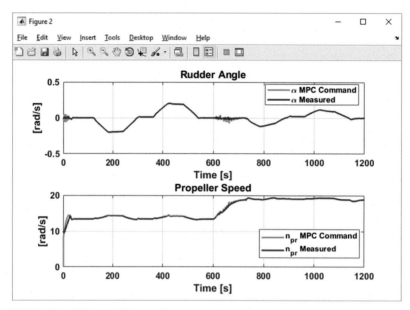

FIGURE 6.27 Multiple MPC controller commands.

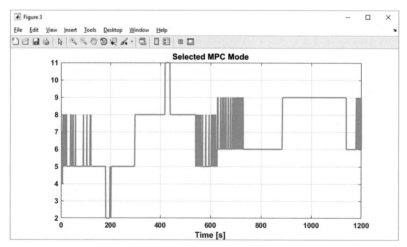

FIGURE 6.28 Multiple MPC controller selected mode.

from one mode to another. Fig. 6.28 shows a high frequency of switching especially at the beginning and middle of the simulation.

To mitigate the high frequency of mode switching, we add a custom-made hysteresis block that will force the controller to stay in the current mode for a minimal period after a change in mode occurs (Fig. 6.29). The main parameter that the designer needs to tune is the duration of the timer.

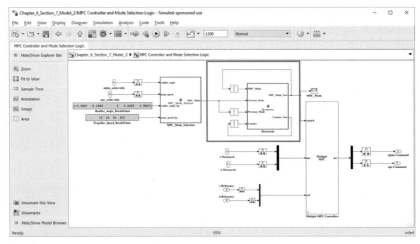

FIGURE 6.29 Hysteresis block for mode selection.

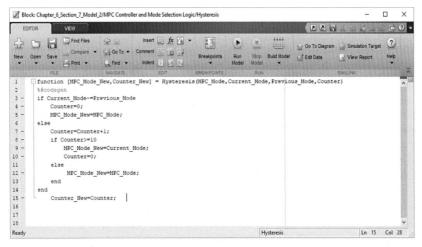

FIGURE 6.30 Hysteresis logic for mode selection.

The longer the timer is, the less the switching will be. The tradeoff is the accuracy of the MPC controller of the previous mode in performing in the current mode. Some iterations might be required to tune the hysteresis block.

Fig. 6.30 shows the hysteresis logic that was implemented. The value for the counter that we chose was 10. As was mentioned in Chapter 5, Single MPC Design for a Ship, the execution rate of the MPC controller is 1 second. This means that the hysteresis block will not send the new mode to the MPC controller until 10 seconds have elapsed, since the mode has changed.

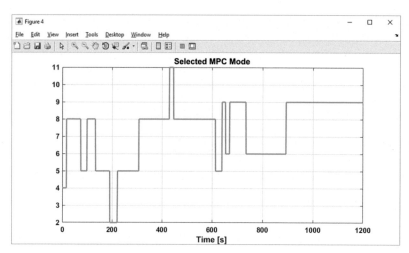

FIGURE 6.31 Multiple MPC controller selected mode.

The updated Simulink model with the hysteresis block, *Chapter_6_Section_8_Model.slx*, can be found in Chapter_7/Section_8. To run the model and plot the results, run *Chapter_6_Section_8_Script_2.m* which can be found in the same folder. Fig. 6.31 shows the selected modes after adding the hysteresis block. Comparing Fig. 6.28 and Fig. 6.31, it is evident that the hysteresis block reduced the switching.

The tracking performance is shown in Fig. 6.32. At the beginning of the simulation, there is a deterioration in the tracking performance of the turning rate as compared to Fig. 6.26. Also, there is bigger overshoot in the forward speed in the case of the multiple MPC with hysteresis. Fig. 6.33 shows the MPC command for the actuators. The rudder command is less chattery than the case with no hysteresis (Fig. 6.27). This is due to the weight of the actuators' rate in the cost function. In the case of hysteresis, a multiple MPC controller is stuck to one mode for 10 seconds each time a mode change is requested. This means that the cost function of the MPC will penalize moving the rudder and propeller. In the case of no hysteresis, every time the mode is switched to a new one, the cost associated to moving the actuators will restart from zero. In the first 20 seconds of Fig. 6.27, the command to the rudder machine was oscillating in a much higher frequency than the turning rate of the rudder. The rudder remained at zero position. Coincidentally, this leads to a better tracking performance at the beginning of the simulation for the case of no hysteresis. As for the tracking performance for the time from 600 to 700 seconds, hysteresis improved tracking, especially for forward speed (compare bottom plot of Figs. 6.26 and 6.32).

Fig. 6.34 shows the comparison of the modes with and without hysteresis in the 2-dimensional actuator space. It can be noticed that in the case without hysteresis, the mode selection ping-pongs between the pair of modes (4 and 7),

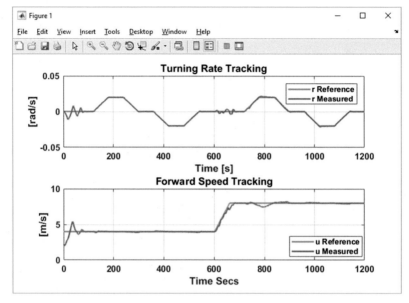

FIGURE 6.32   Multiple MPC with hysteresis controller tracking.

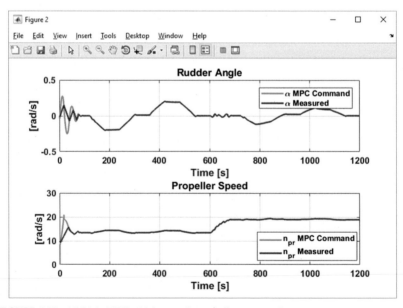

FIGURE 6.33   Multiple MPC with hysteresis controller commands.

(5 and 8), and (6 and 9). In general, adding hysteresis for mode switching is desirable especially if it is properly tuned. It doesn't necessarily yield a better tracking (as we've seen in the beginning of the simulation for the cases with and without hysteresis), but it can lead to less chattery actuator command.

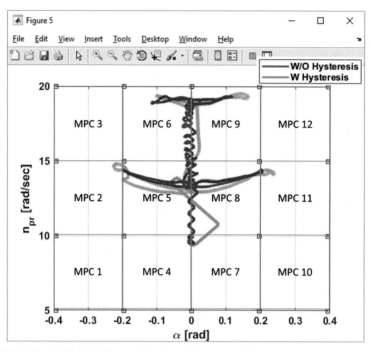

FIGURE 6.34 Multiple MPC controller selected mode with and without hysteresis.

6.9 APPLICATION PROBLEM

In this application, the reader is given the Simulink model for a ship with multimode MPC controllers and a hysteresis block which has a counter of 10 (same model as the one in the previous section). The model's name is *Chapter_6_Section_9_Model.slx* and can be found in *Chapter_6/Section_9*. The script that loads the data for the model and the controller is *Chapter_6_Section_9_Script.m* which can also be found in the same folder.

1. Update the model *Chapter_6_Section_9_Model.slx* to have the hysteresis counter as a constant input to the hysteresis block. The constant variable name is *Counter_threshold*. Save the model.
2. Update the script *Chapter_6_Section_9_Script.m* to sweep the range of *Counter_threshold* from 1 to 12, with increments of 1.
3. Which value of *Counter_threshold* provides the best turning rate tracking?
4. Which value of *Counter_threshold* provides the best forward speed tracking?

The model that has *Counter_threshold* as a constant input to the hysteresis block can be found in *Chapter_6/Section_9* under the name

Chapter_6_Section_9_Solution.slx. The script that runs the sweep of *Counter_threshold* can be found in the same folder under the name *Chapter_6_Section_9_Solution.m.*

REFERENCES

[1] N.G. Chalhoub, N. Khaled, Guidance and control system for under-actuated marine surface ships and other autonomous-platforms, US Patent 9213336.
[2] N. Khaled, N.G. Chalhoub, A self-tuning guidance and control system for marine surface vessels, Nonlinear Dynamics, Vol. 73, Springer, 2013, pp. 897–906. Issue 1–2.

Chapter 7

Monte-Carlo Simulations and Robustness Analysis for a Multiple MPC of a Ship

7.1 INTRODUCTION

The objective of this chapter is to guide the control designer through the process of analyzing the robustness of a controller operating under various uncertainties. The robustness analysis relies on Monte-Carlo simulations for the ship's navigation under various weather conditions.

This chapter builds on Chapter 6 MPC controller. The tuned Multi Mode MPC will be used as a starting point for the robustness analysis. Modifications will be introduced on the MPC tuning and the robustness analysis will be repeated. A final recommendation will be made to decide which set of tuning meets the optimal robustness versus performance trade-off.

The chapter concludes with final remarks and recommended readings for the topic of robustness analysis.

All the codes used in the chapter can be downloaded for free from *MATLAB File Exchange*. Follow the below link and search for the ISBN or title of this book.

https://www.mathworks.com/matlabcentral/fileexchange/

Alternatively, the reader can download the material and other resources from the dedicated website or contact the authors for further help.

https://www.practicalmpc.com

7.2 INTRODUCING UNCERTAINTIES IN WEATHER CONDITIONS

Environmental conditions can significantly affect the navigation of a ship. As discussed in Chapter 4, waves, wind, and current exert unmeasured external forces on the ship that challenge the controller. For Chapters 4, 5, and 6, the weather conditions were intentionally chosen to be mild (Table 7.1). This helps to achieve a more accurate system identification as well as simplifying the tuning process. Note that wind and current angles are chosen to be the same as wave angle.

Practical Design and Application of Model Predictive Control.
DOI: https://doi.org/10.1016/B978-0-12-813918-9.00007-1
© 2018 Elsevier Inc. All rights reserved.

TABLE 7.1 Weather Parameters for Mild Conditions

Parameter	Value	Unit
Wind Speed	5	m/s
Current Speed	1	m/s
Wave Angle	0	radians

TABLE 7.2 Weather Parameters for the Normal Distribution

Parameter	Mean	Standard Deviation	Unit
Wind Speed	20	10	m/s
Current Speed	10	5	m/s
Wave Angle	1	1.2	radians

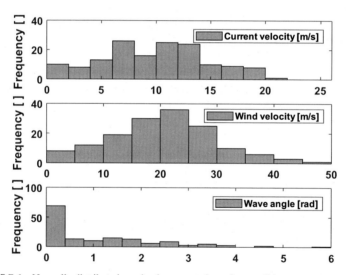

FIGURE 7.1 Normally distributed, randomly generated weather conditions.

Table 7.2 shows the mean and standard deviation of the wind and current speed as well as wave angle that were chosen to simulate random environmental conditions. These values were used to create 150 normally distributed random samples that represent 150 random environmental conditions. These can be generated in MATLAB using the command **normrnd**. If the reader doesn't have MATLAB Statistics and Machine Learning Toolbox, Excel command **NORMINV** with **RAND** can be used. The values below zero were clipped to zero (Fig. 7.1).

To replicate the results in this section, download Chapter 7 material. Under the folder ***Chapter_7/Section_2***, you can run ***Chapter_7_Section_2.m.***

7.3 MONTE-CARLO SIMULATIONS PROCESS

Monte-Carlo is a process of repeated random sampling to obtain numerical results. In controls, it is used to check for the robustness of the control system subject to random uncertainty [1]. The source of uncertainty can be the parameters of the plant, external disturbances, sensing system variability, actuation system variability, etc. In this section, uncertainty is limited to unmeasured external disturbances.

As described in section 7.2, 150 environmental conditions were randomly generated. At each simulation, the ship model and MPC controller are subjected to random weather conditions. To assess the performance of the controller under each condition, we need to select meaningful tracking metrics for the system. We will use the root mean square tracking error for the ship's speed and turning rate. Fig. 7.2 summarizes Monte-Carlo simulation process.

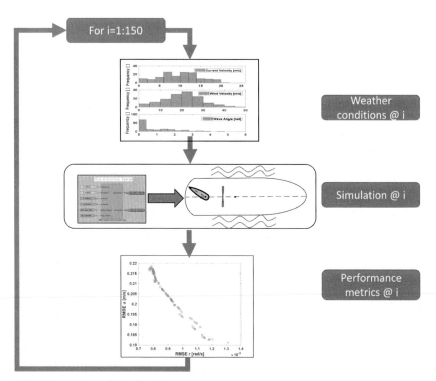

FIGURE 7.2 Monte-Carlo simulation process.

7.4 MONTE-CARLO SIMULATION RESULTS FOR ORIGINAL MPC TUNE

In the previous chapter, the MPC was designed, tuned, and tested under environmental conditions summarized in Table 7.1. Executing the process highlighted in Fig. 7.2, we trace the root mean square tracking error for both turning rate and ship speed (Fig. 7.3). Ideally, the best tuning would yield the smallest mean tracking error, for the 150 simulations, for both turning rate and ship speed and minimal variability in tracking error. We propose to use the below formula (Eq. 7.1) to capture the performance of the controller for the 150 simulations.

$$Penalty = \sqrt[2]{\left(\frac{mean(rmse_r)}{scale_r} \times \frac{std(rmse_r)}{scale_r}\right) + \left(\frac{mean(rmse_u)}{scale_u} \times \frac{std(rmse_u)}{scale_u}\right)}$$

(7.1)

where

$mean(rmse_r)$ is the mean of the root mean square tracking error for the ship's turning rate for the 150 simulations
$std(rmse_r)$ is the standard deviation of the root mean square tracking error for the ship's turning rate for the 150 simulations
$scale_r$ is the range of r (maximum minus minimum value of r) and is equal to 0.04 [rad/s] (refer to Chapter 5 for more details)

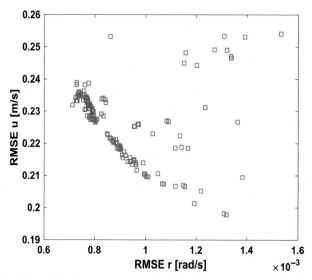

FIGURE 7.3 Monte-Carlo simulation results for original MPC tuning.

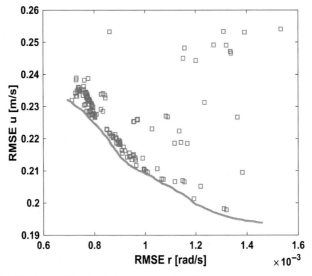

FIGURE 7.4 Pareto front for the root mean square tracking error.

$mean(rmse_u)$ is the mean of the root mean square tracking error for the ship's speed for the 150 simulations

$std(rmse_u)$ is the standard deviation of the root mean square tracking error for the ship's speed for the 150 simulations

$scale_u$ is the range of u (maximum minus minimum value of u) and is equal to 25 [m/s] (refer to Chapter 5 for more details)

Penalty will increase if either the mean or the standard deviation of the root mean square error increases for either of the two outputs. Our objective is to find a tuning that minimizes **Penalty**. Fig. 7.3 shows the results of Monte-Carlo simulations. Fig. 7.4 shows a hand-sketched Pareto Front of the root mean square tracking error. The Pareto Front shows that there is a trade-off between the tracking error of r and that of u. There won't be solutions that exist below the Pareto Front. The value of **Penalty** is 0.0102 for the original tuning of MPC.

To replicate the results in this section, download Chapter 7 material. Under the folder **Chapter_7/Section_4**, you can run **Chapter_7_Section_4.m**.

7.5 IMPACT OF TUNING ON ROBUSTNESS OF MPC

In this section, we will examine how tuning impacts the robustness of the system. We will increase the weight of the output r as compared to u and rerun Monte-Carlo simulation process highlighted in Fig. 7.2. Below are the command lines to change the weights for the 12 modes. Note that the ratio of weight of r to that of u is 2 to 1.

```
MPC_Controller_Mode_1.Weights.OutputVariables = [0.3000*2 0.3000];
MPC_Controller_Mode_2.Weights.OutputVariables = [0.3000*2 0.3000];
MPC_Controller_Mode_3.Weights.OutputVariables = [0.3000*2 0.3000];
...
...
MPC_Controller_Mode_12.Weights.OutputVariables = [0.3000*2 0.3000];
```

Similarly, we will increase the weight of output *u* as compared to *r*. Below are the command lines to change the weights for the 12 modes. Note that the ratio of weight of *r* to that of *u* is 1 to 2.

```
MPC_Controller_Mode_1.Weights.OutputVariables = [ 0.3000 0.3000*2];
MPC_Controller_Mode_2.Weights.OutputVariables = [ 0.3000 0.3000*2];
MPC_Controller_Mode_3.Weights.OutputVariables = [ 0.3000 0.3000*2];
...
...
MPC_Controller_Mode_12.Weights.OutputVariables = [ 0.3000 0.3000*2];
```

Fig. 7.5 shows that in the tuning that had more weight on *r*, tracking deteriorated for *u* (which is expected), but no noticeable improvement in tracking *r* is seen.

As for the tuning that had more weight on *u*, the tracking is worse for both *r* and *u* for all 150 simulations (green circles are to at the top left of the figure).

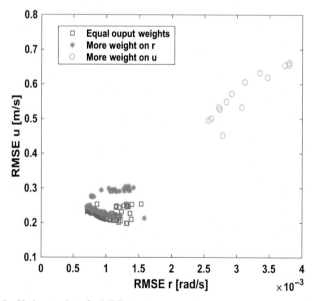

FIGURE 7.5   Various tunings for MPC.

Penalty is 0.0102, 0.0106, and 0.0918 for the equal weights, more weight on ***r***, and more weight on ***u*** respectively. Thus, the original tuning is the most robust tuning.

It is worthwhile to mention that Fig. 7.5 doesn't show all the values of the third tuning. Many of the points were outside the range of the plot to the top right of the figure. The figure was intentionally cropped to zoom in on the data.

To replicate the results in this section, download Chapter 7 material. Under the folder ***Chapter_7/Section_5***, you can run ***Chapter_7_Section_5.m***.

7.6 APPLICATION PROBLEM

The ship that was used in the design of MPC was unloaded. In this section, we explore the robustness of MPC tuning against a 50 percent increase in the mass of the ship as it navigates loaded. Download the model ***Chapter_7/ Section_6/Chapter_7_Multiple_MPC_Loaded.slx***

1. Rerun Monte-Carlo simulations for the three tunings:
   -0.3 for both *r* and *u*
   -0.3x2 for *r* and 0.3 for *u*
   -0.3 for *r* and 0.3x2 for *u*
2. Compare ***Penalty*** of the three tunings with those of section 7.5. Did they all get worse with the increase in uncertainty?

To simulate the solution for this section, download Chapter 7 material. Under the folder ***Chapter_7/Section_6***, you can run ***Chapter_7_Section_6.m***.

REFERENCE

[1] Multivariable Control of Dual Loop EGR Diesel Engine with a Variable Geometry Turbo, Khaled, N., Cunningham, M., Pekar, J., Fuxman, A et al., SAE Technical Paper 2014-01-1357, 2014.

Chapter 8

MPC Design for Photovoltaic Cells

8.1 INTRODUCTION

According to EPA [1], global annual average temperature (as measured over both land and oceans) has increased by more than 0.8°C since 1880 through to 2012. This increase has been attributed to many factors. The monotonic increase in the trends of CO_2 concentration is considered the main factor among them [1]. Photovoltaic (PV) energy has emerged as a potential renewable source of power that significantly competes with the conventional power generation approaches in terms of efficiency and environmental impact.

Improving the efficiency and lifecycle of PV cells is a continuous effort that many research communities are pursuing. Records of National Renewable Energy Laboratory (NREL) [2] indicate that the highest PV cell efficiency is 46% for multijunction (four junctions or more) with concentrators measured at standard test conditions. This efficiency drastically deceases under operational conditions, especially as the test cell temperature increases.

PV module temperature is a significant factor that inversly affects its efficiency. The higher the temperature is, the lower its efficiency is. Passive optical filters are tools used to filter out undesired spectrum wavelengths to reduce their impact on the module temperature. Active optical filters dynamically change the cutoff wavelength based on module temperature and output power during the course of the day. This would help to improve efficiency as well as lifecycle. An optimal active controller for PV applications has been presented in Ref. [3].

In this chapter, MPC will be used to control the active optical filter to achieve optimal output power. The thermoelectrical model [3] is used as a testbed to demonstrate effectiveness. The plant is introduced in Section 8.2. Reference signal generation for the controller is discussed in Section 8.3. The identification and linearization of the plant are discussed in Section 8.4. The physical constraints of the system that are relevant to the controller are discussed in Section 8.5. The detailed process to design a MPC controller for a PV module is covered in Section 8.6. The controller is integrated with the plant in Section 8.7. The chapter is concluded with an analysis of the performance of the controller.

Practical Design and Application of Model Predictive Control.
DOI: https://doi.org/10.1016/B978-0-12-813918-9.00008-3

All the codes used in the chapter can be downloaded from *MATLAB File Exchange*. Follow the below link and search for the ISBN or title of the book.
https://www.mathworks.com/matlabcentral/fileexchange/
Alternatively, the reader can download the material and other resources from the dedicated website or contact the authors for further help.
https://www.practicalmpc.com/

8.2 INTRODUCING THE PHOTOVOLTAIC THERMOELECTRICAL MODEL

PV module efficiency decreases as its temperature increases. Only a portion of the incident irradiance is converted into electrical power. At each wavelength, some energy is reflected—some is absorbed in the PV cells (contributing to electricity production), some is absorbed in other module materials, and some is transmitted through the module. A model capable of predicting this wavelength-specific behavior will generally allow better assessment of the module performance, especially when it is combined with various subsystems such as optical filters.

A wavelength-based thermoelectrical model has been developed [4] and parameterized [5] to predict the temperature and output power of PV modules.

The PV module temperature is a function of the incident radiant power density, the output electrical power, the thermal properties of the materials composing the module, and the heat transfer exchange with surroundings. The main heat transfer paths and energy flow to and from the module are shown in Fig. 8.1.

The rate of change in the module temperature is a function of the incident light which is referred to as shortwave radiation q_{sw}, longwave radiation q_{lw},

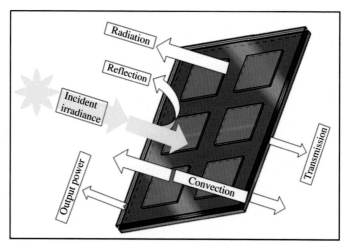

FIGURE 8.1 Heat transfer and energy exchange in the PV module.

heat convection to the surroundings q_{conv}, and output power P_{out}. This can be expressed as:

$$C_{module} \frac{dT_m}{dt} = q_{sw} - q_{lw} - q_{conv} - P_{out}. \qquad (8.1)$$

These components have been derived and discussed in detail [6]. The integrated time domain model is given in Eq. (8.2). Nomenclatures are given in Table 8.1

TABLE 8.1 Nomenclature

Symbol	Definition
C_{module}	Module heat capacity (J/K)
T_m	Module temperature (K)
A	Area of module (m^2)
d_m	Depth of material in module (m)
λ_1	Starting wavelength of spectrum (nm)
λ_2	Ending wavelength of spectrum (nm)
F_{rr}	Solar spectral irradiance (W/(m^{2}nm))
λ	Wavelength (nm)
M	Number of cells in module
A_j	Area of the $p-n$ junction (m^2)
σ	Stefan–Boltzmann const. (5.669 × 10^8 W/(m^{2}K^4))
$\beta_{surface}$	Tilt angle measured with the horizontal (degrees)
ε_{sky}	Emissivity of sky (0.95)
T_{sky}	Effective sky temperature (K)
ε_{gro}	Emissivity of surface of ground (0.95)
T_{gro}	Ground temperature (K)
T_{amb}	Ambient temperature (K)
$h_{c,forced}$	Forced convection coefficient (W/(m^{2}K))
$h_{c,free}$	Free convection coefficient (W/(m^{2}K))
ε_m	Emissivity of PV module (0.9)
α_c	Absorption coefficient of PV cell
α_i	Absorption coefficient of insulation

$$C_{\text{module}} \frac{dT_{\text{m}}}{dt} = \left(\left(\int_{\lambda_1}^{\lambda_2} \alpha_{\text{c}}(\lambda) F_{\text{rr}}(\lambda) d\lambda \right) (MA_{\text{j}}) + \left(\int_{\lambda_1}^{\lambda_2} \alpha_{\text{i}}(\lambda) F_{\text{rr}}(\lambda) d\lambda \right) (A - MA_{\text{j}}) \right)$$

$$- \sigma A \left(-\frac{1 + \cos(\beta_{\text{surface}})}{2} \varepsilon_{\text{sky}} T_{\text{sky}}^4 - \frac{1 - \cos(\beta_{\text{surface}})}{2} \varepsilon_{\text{gro}} T_{\text{gro}}^4 + \varepsilon_{\text{m}} T_{\text{m}}^4 \right)$$

$$- \left(\sqrt[3]{h_{\text{c,free}}^3 + h_{\text{c,forced}}^3} \right) A(T_{\text{m}} - T_{\text{amb}}) - MVI$$

$$(8.2)$$

The purpose of this model is to predict the effect of each wavelength of the module on the PV module temperature and the output power. As the output power is affected by the module temperature, this interaction requires a controller that finds the optimal cutoff spectral wavelength to minimize the module temperature while maximizing the output power over a period of time. Therefore, the objective of the MPC controller is to maximize the output power by controlling the input power through filtering the spectrum wavelengths.

8.3   CONTROLLER REFERENCE GENERATION

Ideally, the output power of the PV module is the signal that the controller should be. The maximum output power signal is a function of ambient temperature and PV module temperature as given in Eq. (8.2). We don't have the optimal output power under all ambient conditions. Therefore, the module temperature will be used as the controlled signal that is selected to yield optimal output power at a given ambient temperature (details to follow below). The block diagram for the reference generator, controller, and the plant is given in Fig. 8.2.

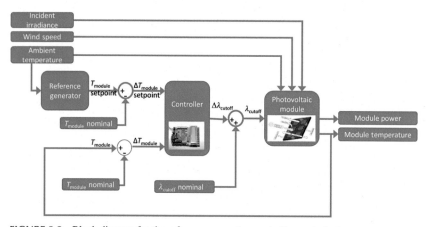

FIGURE 8.2   Block diagram for the reference generator, controller, and plant.

To find the optimal output power at each ambient temperature, we run a design of experiment (DOE). The DOE has two factors: Ambient temperature and ultraviolet (UV) cutoff wavelength λ_{UV}. The factors of the DOE will assume the following levels:

1. The UV cutoff wavelength range is $300 \leq \lambda_{UV} \leq 430$ divided into 14 levels (a step of 10 nm each level).
2. The ambient temperature range is $280 \leq T_{amb} \leq 340$ divided into 10 levels (a step of 6.67 °K each level).

Both the PV temperature and the output power will be recorded at each simulation. The duration of the simulation needs to be determined based on the time constant of the system. For the plant under consideration, it was found that 1800 seconds (30 minutes) is a good time for reaching a steady state for step inputs. The total number of the simulations is 140 and each simulation is 1800 seconds long.

The Simulink model used for the DOE is shown in Fig. 8.3. The inputs to the model are the ultraviolet wavelength, λ_{UV}, the infrared wavelength,

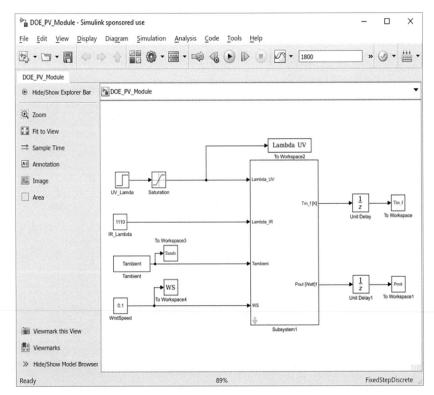

FIGURE 8.3 Simulink model used for the DOE.

λ_{IR}(constant and is equal to 11,100 nm), the ambient temperature, T_{amb}, and the wind speed, WS. λ_{UV} is the manipulated variable. T_{amb} and W_{spd} are measured inputs. The PV model calculates the input power and heat transfer as per Eqs. (8.1) and (8.2). The outputs of the PV plant are output power and the module temperature—both are measurable.

Fig. 8.4 shows the PV module output power versus ambient temperature in the top plot. The bottom plot of Fig. 8.4 shows the PV module temperature versus ambient temperature. The maximum output power is selected at each ambient temperature. Now we can generate a linear fit (Eq. 8.3) for the PV module temperatures and ambient temperature that maximizes the power output. Eq. (8.3) will be used to generate the reference PV module temperature for MPC. The controller will try to maintain this reference temperature, by manipulating cutoff wavelength λ_{UV}, to generate the maximum output power at any given ambient temperature.

$$T_m = 0.916\,T_{amb} + 79.79 \qquad (8.3)$$

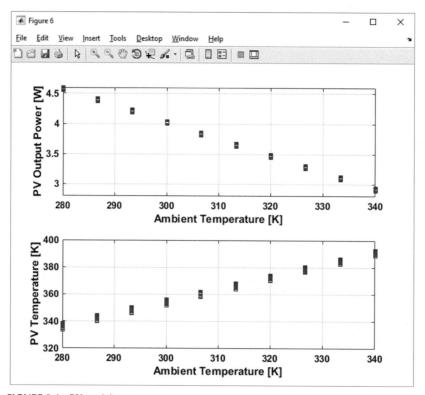

FIGURE 8.4   PV module output power and temperature versus ambient temperature.

The Simulink model in Fig. 8.3 *DOE_PV_Module.slx* and the script to generate Fig. 8.4 *Chapter_8_section_3.m* can be download from *MATLAB File Exchange*. Follow the link below and search for the ISBN or title of the book.

https://www.mathworks.com/matlabcentral/fileexchange/

The material for Chapter 8 can be found under the Chapter_8 folder.

8.4 SYSTEM IDENTIFICATION FOR PHOTOVOLTAIC MODULE

As discussed in Chapter 4, System Identification for a Ship, the first step towards designing a MPC controller is to generate an approximate linearized model for the plant. An experiment will be conducted to collect the input/output data that is needed for the identification. As mentioned above, the thermoelectrical model given in Eq. (8.1) is used as the plant that data will be extracted from. To generate the data needed for linearization, the plant model is excited with two step changes in λ_{UV} (Fig. 8.5). The value of λ_{UV} is changed from minimum (300 nm) to maximum (430 nm) and then

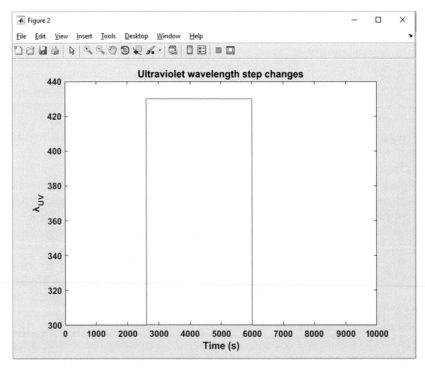

FIGURE 8.5 Ultraviolet wavelength step changes.

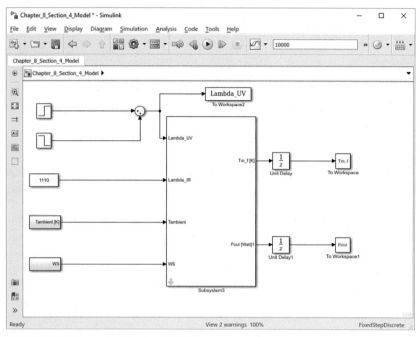

FIGURE 8.6 System identification simulation setup.

back to minimum (300 nm). Fig. 8.6 shows the Simulink model *Chapter_8_Section_4_Model.slx* that is used to generate the data. The manipulated input λ_{IR} is kept constant at 1100 nm. The ambient temperature and wind speed inputs are provided as a lookup table as these inputs change dynamically with time. Changing the λ_{UV} to a value that is higher than 300 nm, would reduce the input power to the module, and accordingly the module temperature and output power would dynamically be impacted. Due to this step change, it takes about 30−45 minutes to reach the steady-state response. Stepping λ_{UV} from 430 to 300 nm increases the input power which increases the module temperature. The overall simulation time is selected to be 10,000 s to allow the PV module enough time to reach steady state.

The m-file code titled *Chapter_8_Section_4_Script.m* performs the system identification. It runs the Simulink model *Chapter_8_Section_4_Model. slx*. Then it clips the simulation data to get rid of the initialization portion of the simulation (the first 2000 data points are clipped). The function *iddata* is used to prepare the data in the proper format for system identification. The system identification function *n4sid* is used to generate an initial linear model. The resultant linearized model is further refined using *pem* function.

To determine the goodness of the linear model, the function *compare* is used. It generates a comparison plot between the actual data and the linear model as shown in Fig. 8.7. *ss* command is used to represent the results from sys2 in a state space format.

Fig. 8.8 shows the resulting state space representation of the linearized model. Note that the NRMSE is 92.36%, which indicates a good match between the actual plant and the linear model. The linear model is saved as *Chapter_8_Section_4_Linear_Model.mat* and will be used later in Section 8.6.

Below is the MATLAB code that the performs the system identification. This code can be found in the *Chapter_8/Section_4* folder under the name *Chapter_8_Section_4_Script.m*

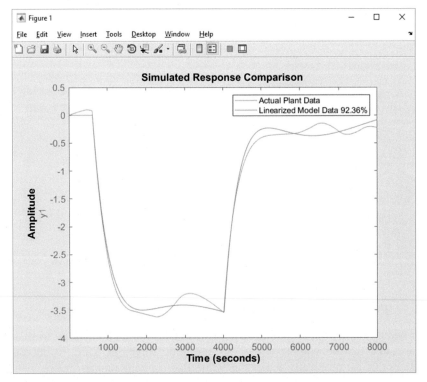

FIGURE 8.7   Responses of the actual and linearized data for step changes.

```
%Book Title: Practical Design and Application of Linear MPC
%Chapter: 8
%Section: 4
%Authors: Nassim Khaled and Bibin Pattel
%Last Modified: 1/16/2018


close all
%% System Identification of PV Cell

%% Simulate the model
sim('Chapter_8_Section_4_Model')
%% Create time vector
Time=0:1:length(Lambda_UV)-1;

%% Change plotting settings
set(0,'DefaultLineLineWidth',1)
set(0,'DefaultAxesFontWeight','bold')
set(0,'DefaultAxesFontName','Arial')
set(0,'DefaultAxesFontSize',12)

%% Grouping simulation data and neglecting first few seconds(Nstart*stp_sz)
Nstart =2000; %Clip the data starting from Nstart as the first few seconds of the simulation
should be discarded
Measured_Outputs=[  Tm_f(Nstart:end)]; %Measured outputs
Manipipulated_Variables=[ Lambda_UV(Nstart:end) ];% Pout_previous ], %Manipulated variables

%% Obtaining initial conditions at step time=Nstart
Measured_Outputs_Nstart=Measured_Outputs(1,:); %Capturing the measured outputs at step time=
Nstart
Manipipulated_Variables_Nstart=Manipipulated_Variables(1,:); %Capturing the manipulated
variables at step time= Nstart

%% Forcing response to start from zero initial conditions
Measured_Outputs_zero_initial_conditions=Measured_Outputs(1:end,:)-
repmat(Measured_Outputs_Nstart,length(Measured_Outputs(1:end,:)),1); %Subtracting initial
conditions for measured outputs to obtain zero response at step time= Nstart
Manipipulated_Variables_zero_initial_conditions=Manipipulated_Variables(1:end,:)-
repmat(Manipipulated_Variables_Nstart,length(Manipipulated_Variables(1:end,:)),1); %Subtracting
initial conditions for manipulated variables to obtain zero actuation at step time= Nstart

%% Prepare date for system identification
stp_sz=10
data=iddata(Measured_Outputs_zero_initial_conditions(1:stp_sz:end,:),Manipipulated_Variables_
zero_initial_conditions(1:stp_sz:end,:),stp_sz); %data is packaged for system identification
using iddata

%% Generate a preliminary 3rd order system that fits the data
sys1=n4sid(data,3,'Form','canonical','DisturbanceModel','None','InputDelay',[0]',
'InitialState','zero'); %n4sid generates a preliminary system in the canonical form

%% Generate a more refined system
sys2 = pem(data,sys1,'InitialState','zero') %pem generates a more refined system that fits the
data better

%% Define the options for comparing the various identified systems
opt = compareOptions('InitialCondition','Z');

%% Plot the comparison between the simulation data and the identified system sys2
figure
compare(data,sys2,opt);

%% Extract A, B, C and D matrices from sys2
A=sys2.A
B=sys2.B
C=sys2.C
D=sys2.D

%% Use the ss command to represent the results from sys2 in a state space format
PV_Linear_Model  = ss(A,B,C,D,stp_sz)
save Chapter_8_Section_4_Linear_Model.matPV_Linear_Model
```

8.5   PHYSICAL CONSTRAINTS OF THE SYSTEM

Providing MPC with the constraints for the actuators, their rates, as well as the range for the outputs allows the quadratic solver of MPC to choose values that are physically achievable while also reducing tuning time.

```
sys =

  A =
              x1        x2        x3
     x1        0         1         0
     x2        0         0         1
     x3     0.9681    -2.936    2.968


  B =
                      u1
     x1     -0.0007632
     x2     -0.0007444
     x3      -0.000726


  C =
           x1   x2   x3
     y1    1    0    0


  D =
           u1
     y1    0
```

FIGURE 8.8   State space representation of the linearized model.

```
Sample time: 10 seconds
Discrete-time state-space model.
```

The range of λ_{UV} is [300, 430]nm. The range for T_m is not fully known as it depends on wind speed, solar conditions, and ambient temperature in addition to λ_{UV}. We will choose a wide range of [262,482]F.

We will design the controller to provide a change in frequency $\Delta\lambda_{UV}$ and not an absolute value of λ_{UV}. This is a common practice in control design. The nominal value for λ_{UV} is 300 nm. The range of $\Delta\lambda_{UV}$ is [0, 130]nm. The nominal value for T_m is 372F. The range of ΔT_m is [− 100, 100]F. The nominal values for λ_{UV} and T_m were obtained from the system identification.

8.6   DESIGNING A MPC CONTROLLER FOR THE PV MODULE

Using the same process introduced in the previous chapter, we design MPC to control the temperature of the PV module.

1. To start the design process, download and load the linear model of the PV module, *Chapter_8_Section_4_Linear_Model.mat*. This model was developed in Section 8.4 and can be found in sections 4 and 6 folders.
2. In the MATLAB command window, type *mpcDesigner* to open *MPC Designer App* GUI. Click on the *MPC Structure* button to configure the controller (Fig. 8.9).

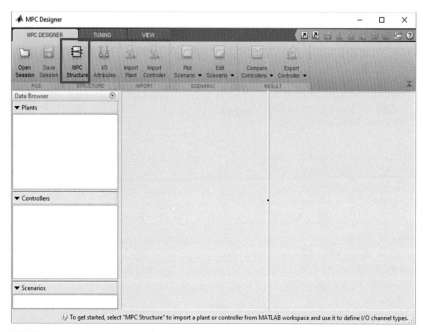

FIGURE 8.9 *MPC Structure* button.

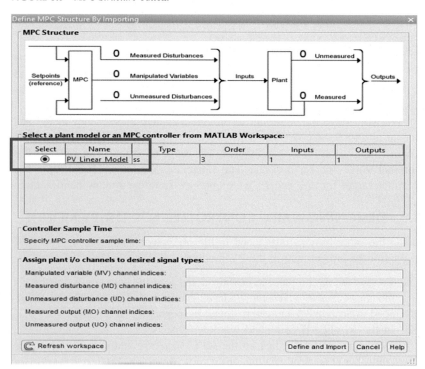

FIGURE 8.10 Selecting the plant model.

3. Select the plant model (*PV_Linear_Model* in Fig. 8.10). You can click on the model to preview the structure (Fig. 8.11).
4. Close the preview of the linear model and click *Define and Import* (Fig. 8.12). You might need to wait for a minute until the default MPC controller is generated. Fig. 8.13 shows what the MPC default setup should like if all the previous steps have been executed correctly.
5. Click the *I/O Attributes* button to define the names, units, scaling factors, and nominal values of the inputs and outputs of the controller (Fig. 8.14).
6. Define the units of the inputs and outputs as indicated in Fig. 8.15. u1 is the change in the cutoff frequency, $\Delta\lambda_{UV}$ (nominal λ_{UV} is 300 nm). y_1 is the change in temperature (nominal temperature is 300K). This step is optional as the units are not necessary for the design of MPC.
7. The MPC Designer App requires that the inputs and outputs to be scaled. Using the open-loop information about the inputs and outputs of the system provided in the previous section, you can fill in the *Nominal* and *ScaleFactor* for the MPC controller as shown in Fig. 8.16. *ScaleFactor* is the set of ranges for the inputs and outputs that were

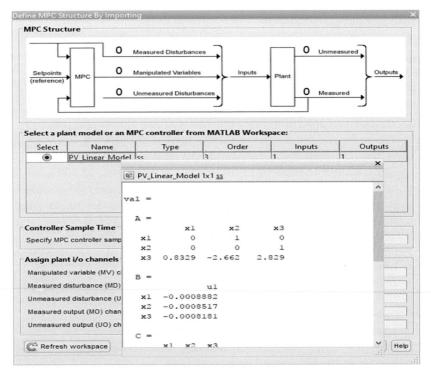

FIGURE 8.11 Preview of the linear model.

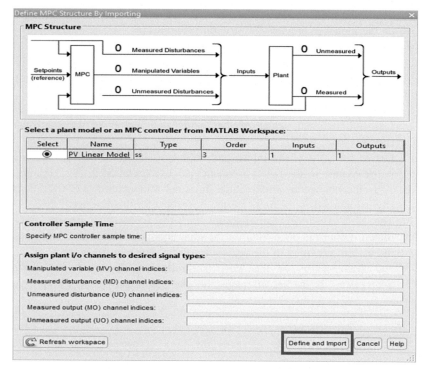

FIGURE 8.12 Define and Import button.

introduced in Section 8.5. Once done entering the values, click *Apply* and then *OK*.

8. To start the tuning of the controller, click on the *Tuning* tab (Fig. 8.17).
9. To choose the control and prediction horizon for MPC, it is important to know the time constant as well as the sampling frequency of the sensors. The time constant of the PV module is in the order of 100 seconds and can be easily obtained by running a simulation where the module is subjected to a step change in λ_{UV}. If the prediction horizon is much smaller than the time constant of the system, there will be a noticeable degradation in the performance of MPC. The temperature sensor sampling frequency is the same as the control action execution frequency, which is 10 seconds. The control horizon was chosen to be 20 seconds. Before setting these values in the tool, it is worthwhile noting that the *Control interval* in the tool is a function of the *Sample time* for the discrete model for the MPC controller. In this case the sample time was 10 seconds. Setting the *Control Interval* to 2 means that the *Control interval* $= 10 \times 2 = 20$ seconds. Set the *Prediction horizon* and *Control horizon* to 10 and 2 respectively (Fig. 8.18).

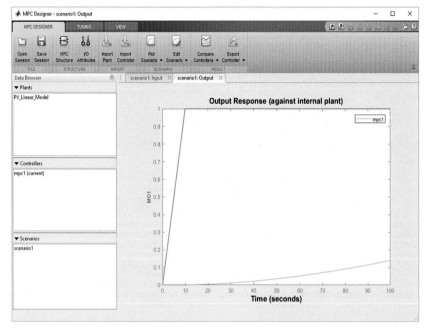

FIGURE 8.13   Default MPC controller.

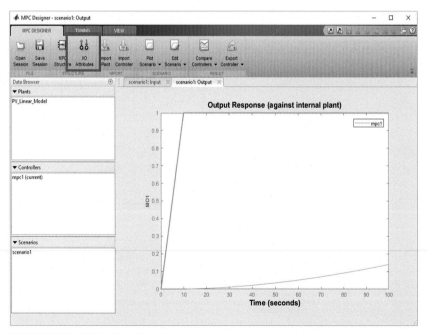

FIGURE 8.14   *I/O Attributes* button.

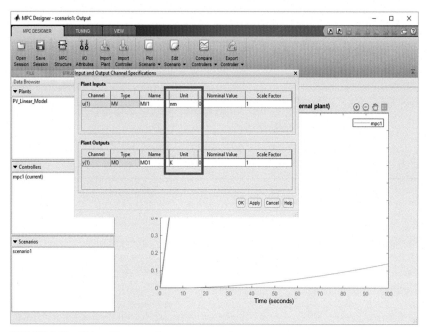

FIGURE 8.15   Units of inputs and outputs.

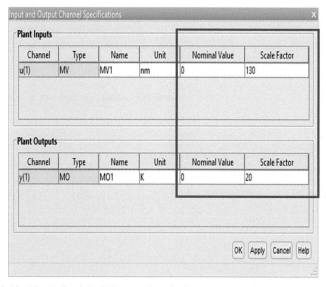

FIGURE 8.16   *Nominal* and *ScaleFactor* values for inputs and outputs.

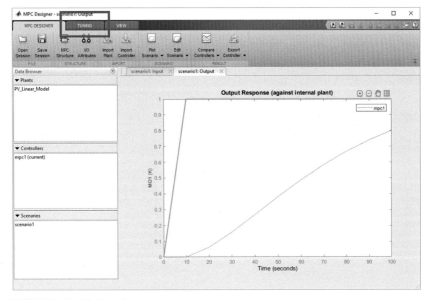

FIGURE 8.17   *Tuning* tab.

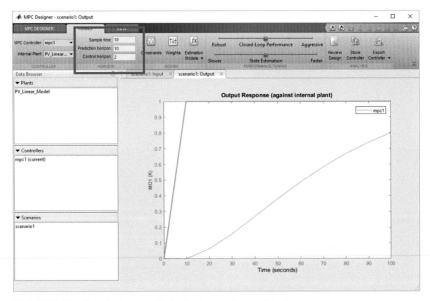

FIGURE 8.18   Sample time, prediction horizon and control horizon.

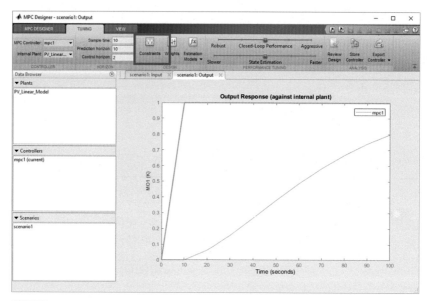

FIGURE 8.19 *Constraints* button.

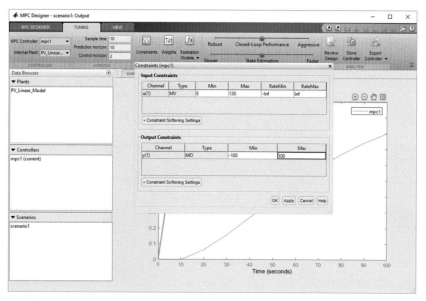

FIGURE 8.20 *Constraints* setup.

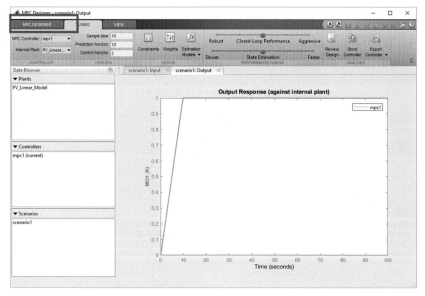

FIGURE 8.21 *MPC Designer* button.

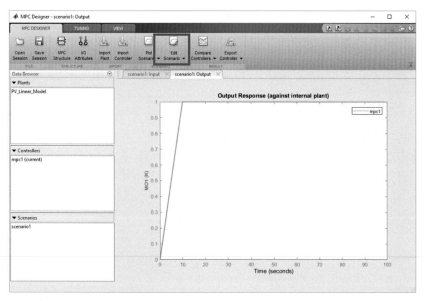

FIGURE 8.22 *Edit Scenario* button.

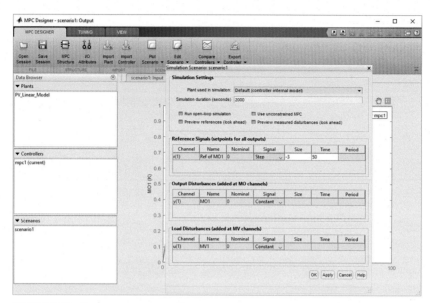

FIGURE 8.23   Setting up a step scenario.

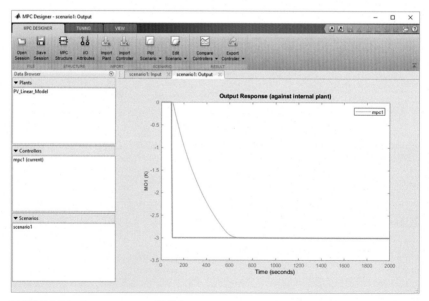

FIGURE 8.24   Output response of MPC.

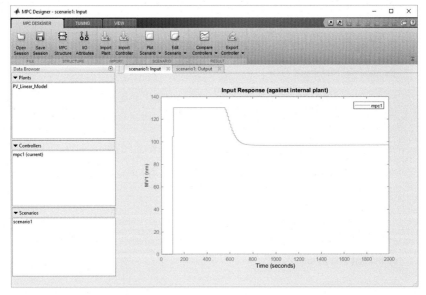

FIGURE 8.25 Input command of MPC.

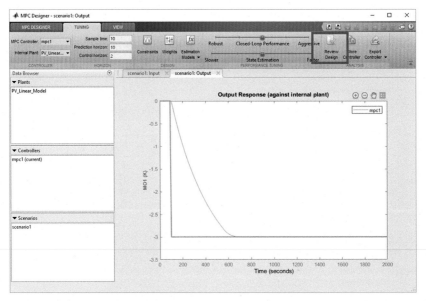

FIGURE 8.26 *Review Design* button.

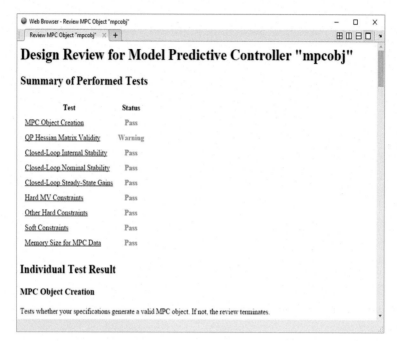

FIGURE 8.27   *MPC design* report.

FIGURE 8.28   QP Hessian matrix validity constraints.

10. Click on the *Constraints* tab to define the range of the actuators, their rates, and the range of the outputs (Fig. 8.19).
11. During execution, MPC will select values for $\Delta\lambda_{UV}$ that respect $\Delta\lambda_{UV}$ minimum/maximum, $\Delta\lambda_{UV}$ rate minimum/maximum, as well as output temperature minimum/maximum. Using the physical constraints from Section 8.5, input the minimum and maximum values. It is worthwhile noting that the $\Delta\lambda_{UV}$ rate doesn't have any constraints and thus we can leave the range as the default which is]-inf,inf[. Fig. 8.20 shows the final setup of the constraints. Press *Apply* and then *OK*. You might need to wait for a minute for these changes to apply.
12. Click on MPC Designer (Fig. 8.21).
13. Click on *Edit Scenarios > scenario1* to setup a step response simulation to challenge the tracking capability for the designed MPC (Fig. 8.22).
14. We will set the simulation duration to 2000 seconds. The reference for the controller is selected to be a step (under *Type*). The step value is −3 (nm) at time 50 seconds (Fig. 8.23). Click *Apply* and *OK*.
15. Fig. 8.24 shows the good tracking response of MPC (scenario 1: Output tab). Fig. 8.25 shows the command of $\Delta\lambda_{UV}$ (scenario 1: Input tab).

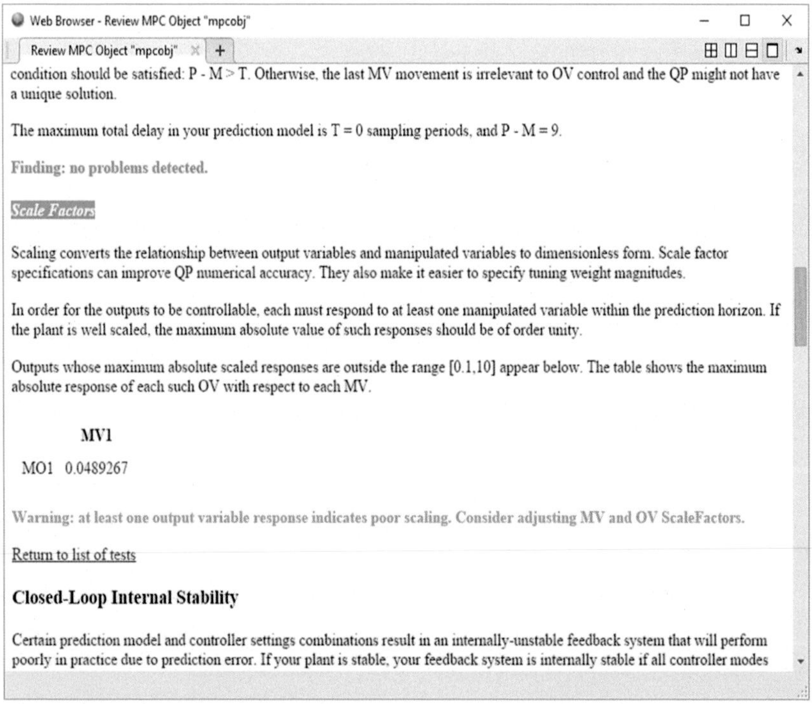
Web Browser - Review MPC Object "mpcobj"

Review MPC Object "mpcobj"

condition should be satisfied: P - M > T. Otherwise, the last MV movement is irrelevant to OV control and the QP might not have a unique solution.

The maximum total delay in your prediction model is T = 0 sampling periods, and P - M = 9.

Finding: no problems detected.

Scale Factors

Scaling converts the relationship between output variables and manipulated variables to dimensionless form. Scale factor specifications can improve QP numerical accuracy. They also make it easier to specify tuning weight magnitudes.

In order for the outputs to be controllable, each must respond to at least one manipulated variable within the prediction horizon. If the plant is well scaled, the maximum absolute value of such responses should be of order unity.

Outputs whose maximum absolute scaled responses are outside the range [0.1,10] appear below. The table shows the maximum absolute response of each such OV with respect to each MV.

	MV1
MO1	0.0489267

Warning: at least one output variable response indicates poor scaling. Consider adjusting MV and OV ScaleFactors.

Return to list of tests

Closed-Loop Internal Stability

Certain prediction model and controller settings combinations result in an internally-unstable feedback system that will perform poorly in practice due to prediction error. If your plant is stable, your feedback system is internally stable if all controller modes

FIGURE 8.29 *Scale Factors* for manipulated variables.

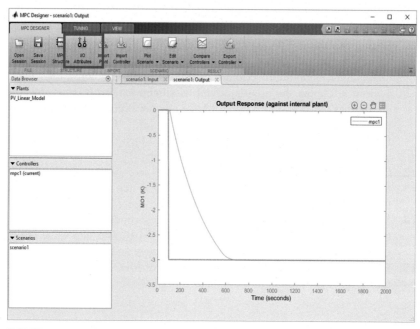

FIGURE 8.30 I/O attributes.

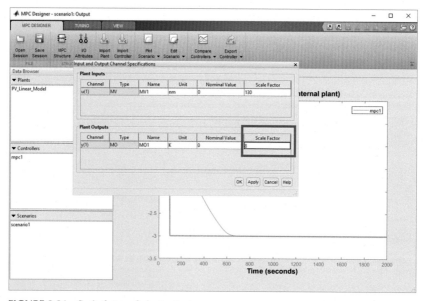

FIGURE 8.31 Scale factor of plant output.

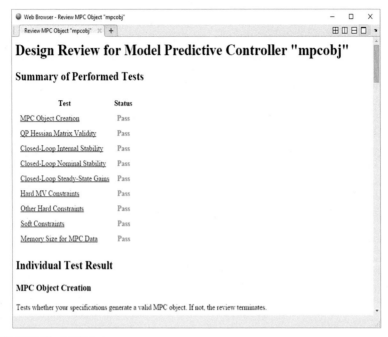

FIGURE 8.32   *MPC design* report.

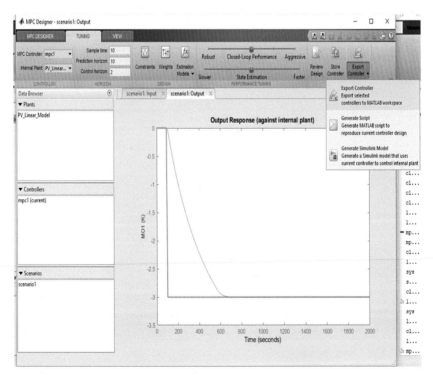

FIGURE 8.33   Export controller button.

Note that the MPC design tool uses the linear PV model combined with the MPC controller to simulate the design.

16. One final step before we can export the MPC controller is to check the health of the MPC design. This is done by clicking on the tuning tab and then clicking on *Analysis > Review Design* (Fig. 8.26). You might need to wait for a minute before the report is generated.

17. Fig. 8.27 shows that there is one warning in orange related to *QP Hessian Matrix Validity*. Click on the hyperlinked text to get a detailed explanation and the recommended action.

18. Fig. 8.28 shows the QP Hessian Matrix Validity. Scroll down to the section titled Scale Factors (Fig. 8.29). The issue with the design is highlighted in orange "Warning: at least one output variable response indicates poor scaling. Consider adjusting MV and OV ScaleFactors." This means that the range we provided for either the input or output variables was not set up properly. The scale factor that we provided for the input and output was 130 nm and 20F respectively.

19. Since we are confident about the range of the input λ_{UV}, we will change *Scale Factor* for the output from 20 to 8. To do so, click on I/O

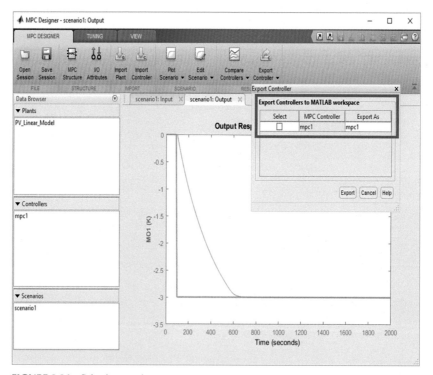

FIGURE 8.34 Selecting mpc1.

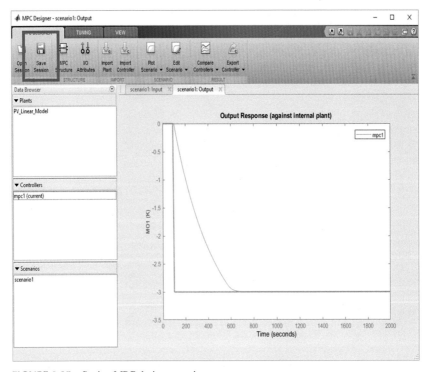

FIGURE 8.35 Saving MPC designer session.

Attributes first (Fig. 8.30). Then change *Scale Factor* of the *Plant Outputs* to 8. Click *OK* once done (Fig. 8.31).

20. Review the MPC design by clicking on the *TUNING* tab and then on *Review Design*. Fig. 8.32 shows that the designed MPC controller passed all the checks.

21. Click *Export Controller* button and select *Export Controller* (Fig. 8.33).

22. Tick the box to select mpc1 in the small pop-up window, then click Export (Fig. 8.34). The controller will be saved to *Workspace* under the name that appears in the *MPC Designer* on the left (*mpc1* in this case).

23. It is a good practice to save the session in case you need to go back and redesign or retune MPC. Click on *Save Session* under *MPC Designer* and save the session (Fig. 8.35). MATLAB will save the session in a .*mat* format that can be loaded in future sessions.

The above session can be downloaded and loaded for verification. It is saved under *MPC_DesignTask_Chapter_8_Section_6.mat*

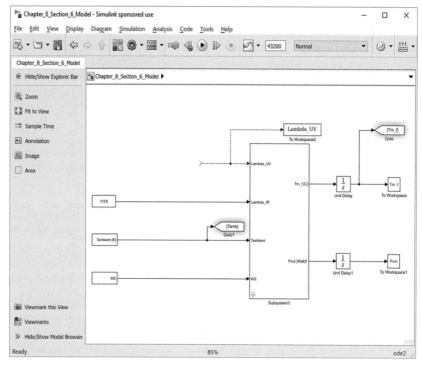

FIGURE 8.36 PV model.

8.7 INTEGRATING MPC WITH THE SIMULINK MODEL

In this section, the designed MPC will be integrated with Simulink for evaluating the designed controller. The model *Chapter_8_Section_7_Model.slx* (Fig. 8.36) will be used as a test platform. It has the PV model, the ambient conditions as well as *To Workspace* blocks setup to send simulation data to workspace. It can be found on the MATLAB Central (File Exchange): https://www.mathworks.com/matlabcentral/fileexchange/

Search for the book title (or ISBN), download Chapter 8/Section 7 material and set as the working directory in MATLAB as the model will load the necessary *.mat* file (*PV_model_data.mat*).

In case you were not able to properly design the MPC controller in the previous section, you will need to load the MPC controller *Chapter_8_Section_7_MPC_Controller.mat*. To build the controller setup of Fig. 8.3, follow the steps below to integrate the controller with the PV model.

1. Add a *MPC Controller* block to the Chapter_8_Section_7_Model.slx model. The block can be found in *Model Predictive Control Toolbox* (Fig. 8.37).

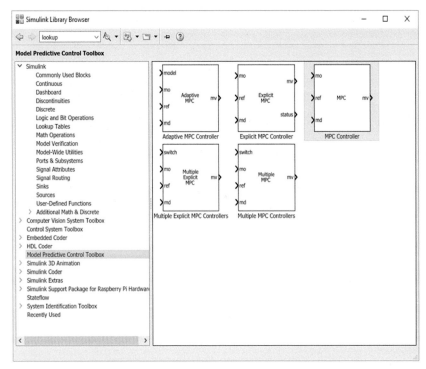

FIGURE 8.37 *MPC Controller* block.

2. Double click the *MPC Controller* block. Type *mpc1* in the parameter *MPC Controller*. Since there is no unmeasured disturbance in the model, untick the *Measured disturbance (md)* (Fig. 8.38). Click *OK* when done.
3. To add the input signal to the controller, add a *From* block and link it to *Tm_f* signal (which is the module temperature) (Fig. 8.39).
4. Add a *Unit Delay* block to break the algebraic loop (Fig. 8.40). Set the initial conditions to 380F (the initial value for the PV module is 380F).
5. Add a *Constant* and a *Sum* block to subtract the PV module temperature from the nominal value, which is 372.8F (from the system identification section) (Fig. 8.41).
6. To generate the reference signal for the controller from ambient temperature, add a *From* block and link it to the *Tamb* signal (which is the ambient temperature) (Fig. 8.42).
7. To implement Eq. (8.3), add a *Fcn* block (highlighted in Fig. 8.43).
8. Change the equation in the *Fcn* block to $0.916 * u(1) + 79$ (Fig. 8.44).
9. As for step 5, add a *Constant* and a *Sum* block to subtract the ambient temperature from the nominal value, which is 372.8F (Fig. 8.45).

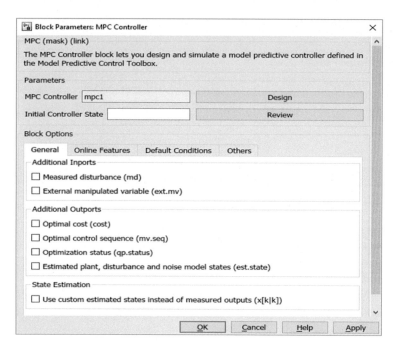

FIGURE 8.38   Embedding the controller in *MPC Controller* block edit.

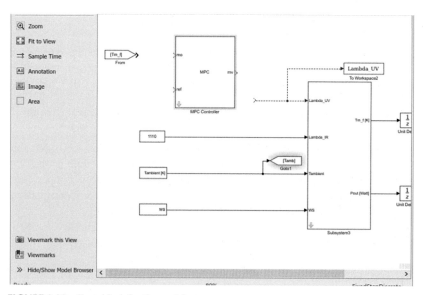

FIGURE 8.39   *From* block for the module temperature.

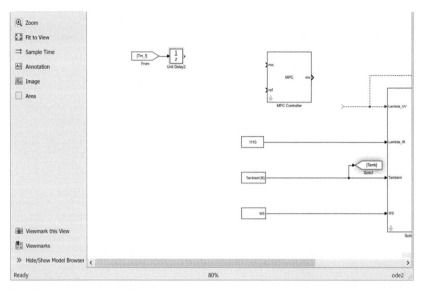

FIGURE 8.40 *Unit Delay* block.

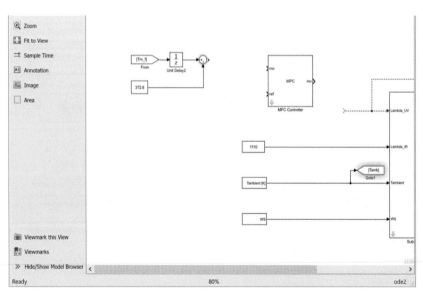

FIGURE 8.41 *Constant* and *Sum* blocks for the PV module temperature.

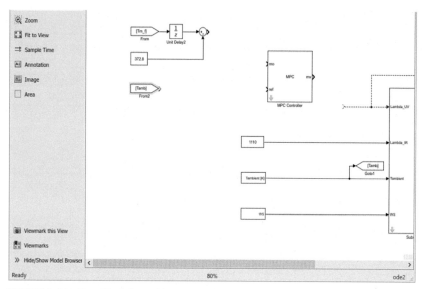

FIGURE 8.42 *From* block for ambient temperature.

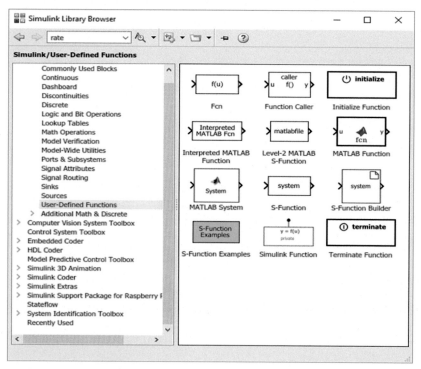

FIGURE 8.43 *Fcn* block.

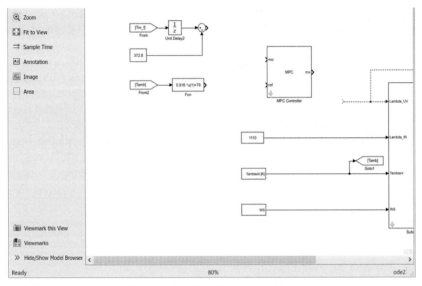

FIGURE 8.44 *Fcn* block with the PV module reference equation.

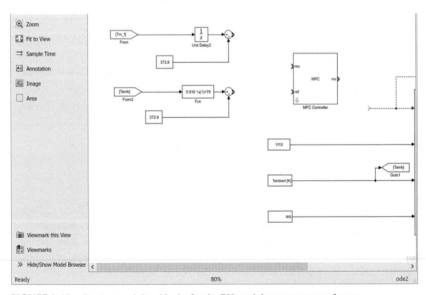

FIGURE 8.45 *Constant* and *Sum* blocks for the PV module temperature reference.

10. To save the generated PV temperature reference, add a *To Workspace* block. Name the generated parameter *T_PV_Reference*. Connect the output of the *Fcn* block to this block (Fig. 8.46).
11. Before we connect the measured output and reference to MPC, we need to check if these inputs run at the same execution rates as MPC. The Simulink model runs at 1 second. This means the two temperature sensors, ambient and PV module, run at 1 second. The MPC controller runs at a sample time of 10 seconds (you can check the *Sample time* in the MPC design section). This means that we need to downsample the measurements to 10 seconds. Add two *Rate Transition* blocks (which can be found under *Simulink/Signal Attributes*). Change the *Outport sample time* to 10 seconds in both (Fig. 8.47).
12. Connect the *Rate Transition* blocks as per Fig. 8.48.
13. At the output of MPC block, we need to upsample from 10 to 1 second. Add a Rate Transition block and change the *Outport sample time* to 1 second.
14. Connect the *Rate Transition* blocks as per Fig. 8.49.
15. Recall that the output of MPC is a $\Delta\lambda_{UV}$. Add a constant block to the output of the *Rate transition* block. The nominal value of λ_{UV} is 300 nm(Fig. 8.50).
16. To clean up the Simulink model, select the whole controller model, with the exception of the ambient and module temperature *From* blocks. Three dots will appear either on the left or right side of the selection (Fig. 8.51).
17. Select *Create Subsystem* (Fig. 8.52).
18. Connect the subsystem to the PV model (Fig. 8.53).
19. Run the simulation and check tracking results (Fig. 8.54).

The model and controller can be found in the Chapter_8/Section_7 material under the name *Chapter_8_Section_7_Model_and_Controller.slx*. Results can be reproduced by running the model.

8.8 CONTROLLER PERFORMANCE

In this section, we will zoom into sections of the simulation results reported in Fig. 8.54 and study the performance of MPC. Fig. 8.55 shows the first hour of the simulation. The temperature of the module is lower than the reference in the top plot of Fig. 8.55. MPC commands the lowest λ_{UV} value which is 300 nm. This allows filtering to occur, which in return will yield the highest PV module temperature achievable at the given ambient conditions.

In the second hour of the simulation (Fig. 8.56), the top plot shows that temperature tracking is achieved, with some oscillations. At about 2 hours, power is at the maximum value of this portion of the simulation.

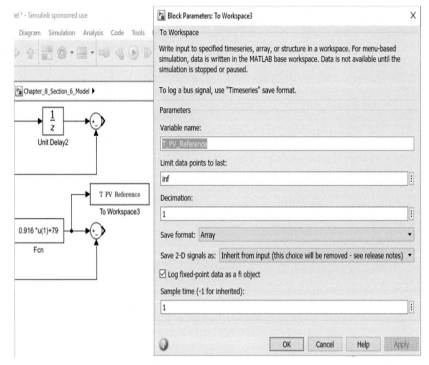

FIGURE 8.46 Setting up *To Workspace* for PV module reference temperature.

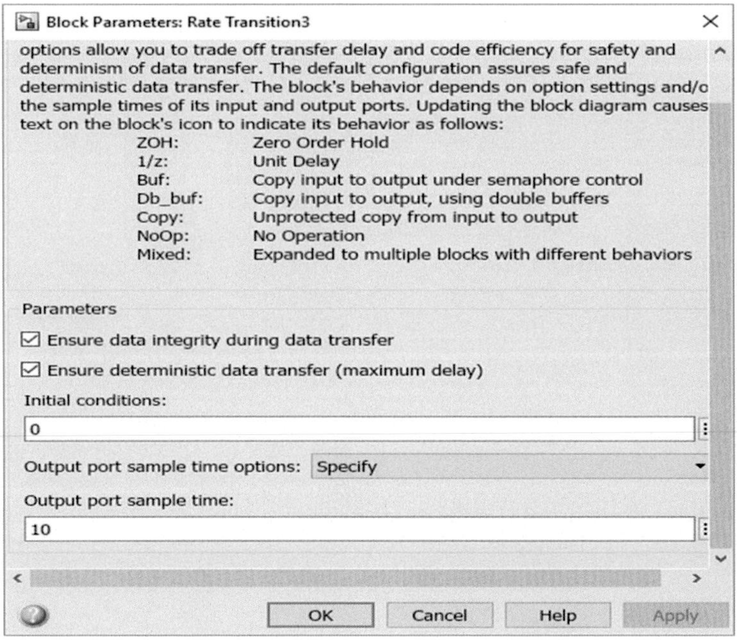

FIGURE 8.47 *Rate transition* block for PV module measured and reference temperatures.

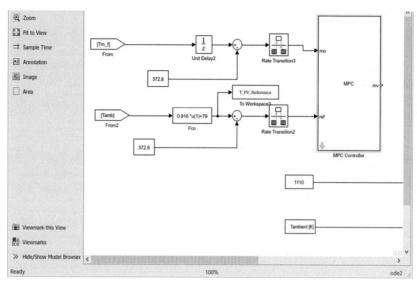

FIGURE 8.48   Connecting *Rate transition* blocks to the inputs of MPC.

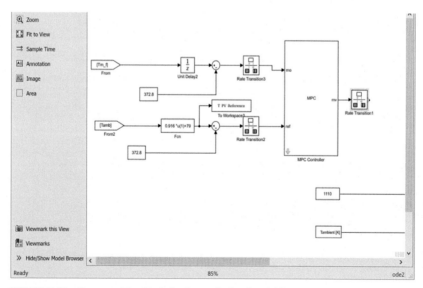

FIGURE 8.49   *Rate transition* block for the manipulated variable.

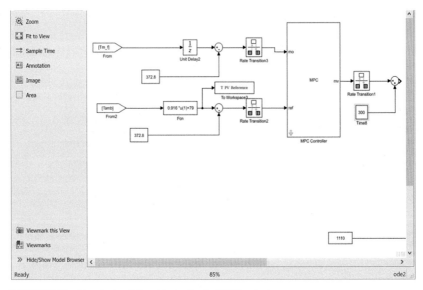

FIGURE 8.50 *Constant* and *Sum* blocks for the MPC output.

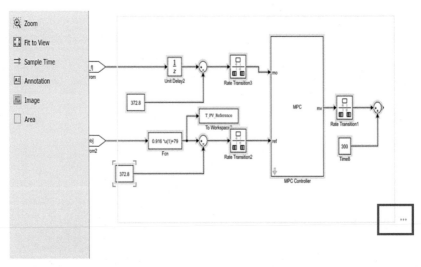

FIGURE 8.51 Selecting the portion of the model to create a subsystem.

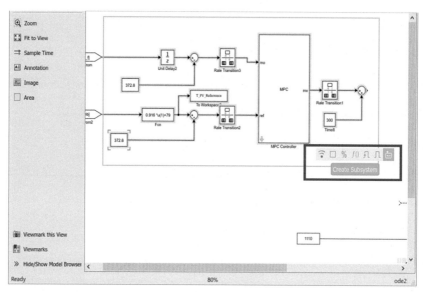

FIGURE 8.52   Creating a subsystem for the MPC controller.

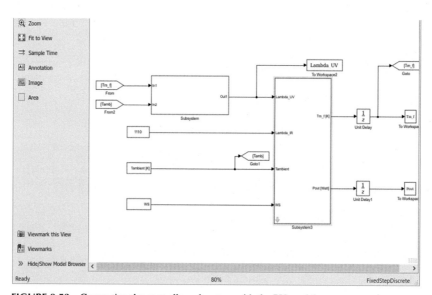

FIGURE 8.53   Connecting the controller subsystem with the PV model.

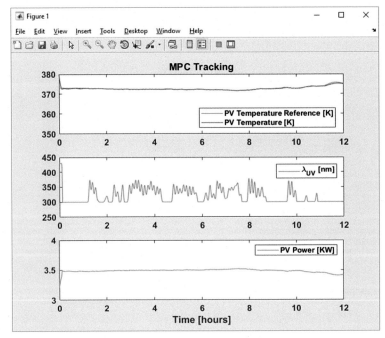

FIGURE 8.54 Simulation results.

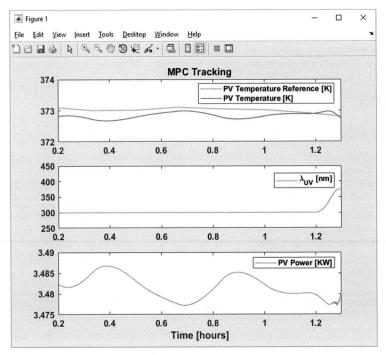

FIGURE 8.55 First hour of the simulation.

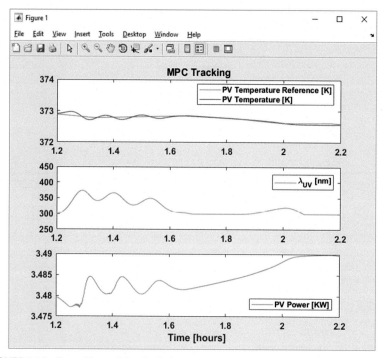

FIGURE 8.56 Second hour of the simulation.

Overall, the tracking performance of MPC is good. But to fully challenge the controller and understand its robustness and benefits, we propose that the reader follows the process outlined in Chapter 7, Monte-Carlo Simulations and Robustness Analysis for a Multiple MPC of a Ship, and run a Monte-Carlo simulation.

REFERENCES

[1] J. Walsh, et al., Climate Change Impacts in the United States, vol. Ch. 2: Our Changing Climate, 2014.
[2] NREL, Best Research-Cell Efficiencies, ed. National Renewable Energy Laboratory, 2017.
[3] S. Aljoaba, A. Cramer, B. Walcott, Active optimal optical filtering of wavelengths for increasing the efficiency of photovoltaic modules, presented at the Photovoltaic Specialist Conference (PVSC), IEEE 40th, Denver, CO, 2014, pp. 1329–1334.
[4] S. Aljoaba, A. Cramer, B. Walcott, Thermoelectrical modeling of wavelength effects on photovoltaic module performance—part I: model, IEEE J. Photovol. 3 (Jul 2013) 1027–1033.
[5] S. Aljoaba, A. Cramer, S. Rawashdeh, B. Walcott, Thermoelectrical modeling of wavelength effects on photovoltaic module performance—part II: parameterization, IEEE J. Photovol. 3 (Jul 2013) 1034–1037.
[6] S. Aljoaba, Active Optimal Control Strategies For Increasing The Efficiency Of Photovoltaic Cells, Ph.D., Department of Electrical and Computer Engineering, University of Kentucky, Lexington, Kentucky, USA, 2013.

Chapter 9

Real Time Embedded Target Application of MPC

9.1 INTRODUCTION

This chapter guides the user through the process of controlling the speed of a DC motor using Model Predictive Control (MPC). The controller is implemented in real-time hardware (Arduino). The System Identification and Controller Design steps are performed similar to previous chapters (Fig. 9.1). The emphasis of this chapter is to outline the process to deploy MPC in the hardware in order to control the motor speed in real time. Technical challenges of embedding the controller are highlighted as well as mitigation plans. Performance of the MPC controller to track a varying speed reference is analyzed in further details.

All the codes used in the chapter can be downloaded from *MATLAB File Exchange*. Follow the link below and search for the ISBN or title of the book.

https://www.mathworks.com/matlabcentral/fileexchange/

Alternatively, the reader can also download the material and other resources from the dedicated website or contact the authors for further help.

https://www.practicalmpc.com/

9.2 CONTROL PROBLEM

The plant model that is being considered is a DC motor. The input to the motor is DC voltage and the output is speed. The objective of the control system is to track a speed reference. To do so, a motor speed sensor provides the MPC controller with a speed reading. The MPC controller generates a PWM command. It is applied to a switching circuit to regulate the DC voltage to the DC motor. Fig. 9.2 shows a block diagram for the closed-loop system.

A block diagram of the control system is shown in Fig. 9.2. Just like any other control system, this block diagram shows a reference generation for the controller to track, a sensing system, feedback controller, actuator system, and a plant. Reference generation generates a reference signal for the motor speed controller to track. The sensing system measures the actual motor

Practical Design and Application of Model Predictive Control.
DOI: https://doi.org/10.1016/B978-0-12-813918-9.00009-5

FIGURE 9.1   Controller design framework.

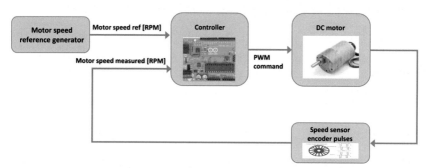

FIGURE 9.2   Motor speed control system block diagram.

speed so that the controller can take necessary corrective action to track the reference speed. The controller in this case is MPC trying to track the reference signal. The controller outputs a PWM signal, which is applied to a switching circuit to regulate the DC voltage to the DC motor, which is the plant to be controlled.

The hardware setup for this follows the work done by University of Michigan CTMS group [1]. The authors used the setup to demonstrate controlling the DC motor using a conventional PI controller. We will study nonlinearity of the plant, and control the setup with single and multiple MPC.

9.3   HARDWARE REQUIREMENTS AND FAMILIARIZATION

Table 9.1 shows all the hardware parts required to replicate the setup described in this chapter. The assembled setup with the MPC controller can also be purchased from the dedicated website of this book https://www.practicalmpc.com/

Arduino Mega is used as a real-time processing platform to run the MPC to control the DC motor in this chapter. The Arduino Mega is a microcontroller board based on the ATmega2560. It simplifies the embedded software deployment and sensor actuator interfaces. We tried using Arduino Uno to run the MPC, but the processing and memory capacity of Arduino Uno is not capable to run the computationally intense MPC controller—mainly because of the program and data memory size. Table 9.2 shows the resources available on the Arduino Mega board.

TABLE 9.1 Hardware Parts Required for Setting up the Example in this Chapter

Item	Description
Arduino Mega	The Arduino Mega is a microcontroller board based on the ATmega2560.
Breadboard and Jumper Wires	Breadboard and Jumper wires are used to make the required electric connection between the hardware parts.
DC Motor with Quadrature Encoder	12 Volt DC motor, with integrated quadrature encoder for speed sensing.
MOSFET (IRF 3205)	MOSFET is used as a switch, controlled by the digital PWM output of the Arduino to connect and disconnect the motor to the battery power source.
Diode (IN4007)	Diode is used to prevent the motor's back EMF causing damage to the circuit.
12V DC Adapter or 12V DC Battery	Either a DC battery or an AC–DC adapter can be used to power the DC motor.

TABLE 9.2 Arduino Mega Board Hardware Specification

Processor	Flash Size	SRAM Size	Digital I/O Pins	PWM Pins	Analog Inputs	Serial Ports
16 MHz ATmega2560	256 kb	8 kb	54	14	16	4

9.4   SIMULINK SUPPORT PACKAGE FOR ARDUINO

MathWorks provides a hardware support package to develop, simulate, and program algorithms, configure and access sensors as well as actuators using Simulink blocks with the Simulink Support Package for Arduino Hardware. Using Simulink external mode interactive simulations, parameter tuning, signal monitoring, and logging can be performed as the algorithms run in real time on the Arduino board.

Follow the below steps to install the Simulink support package for Arduino Hardware.

1. From the MATLAB window go to Home >> Add-Ons >> Get Hardware Support Packages. Fig. 9.3 shows the Add-On Explorer GUI.
2. Click on the Simulink Support Package for Arduino Hardware option, highlighted in Fig. 9.4, and it will guide you to the next window with an

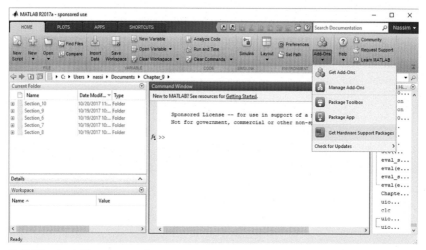

FIGURE 9.3 Add-on explorer GUI.

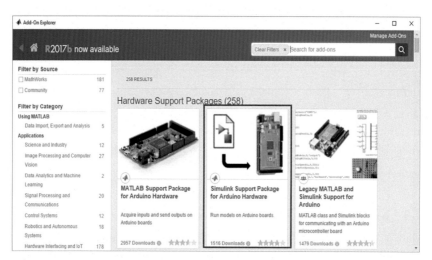

FIGURE 9.4 Simulink support package for Arduino hardware.

Install button, as shown in Fig. 9.5. Click on it. For any support package installation, the user needs to log on to a MathWorks account using the option shown in Fig. 9.6. Log on using the account, or create a new account if you don't have one.

3. Wait for the installation to be completed. An installation progress window is shown in Fig. 9.7. After the successful installation, it will show an option to open the examples available in the support package. Selecting that option will take you to the page with various examples, as shown in Fig. 9.8.

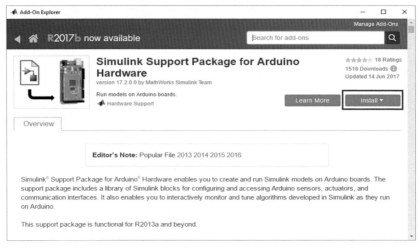

FIGURE 9.5   Add-on explorer with install button.

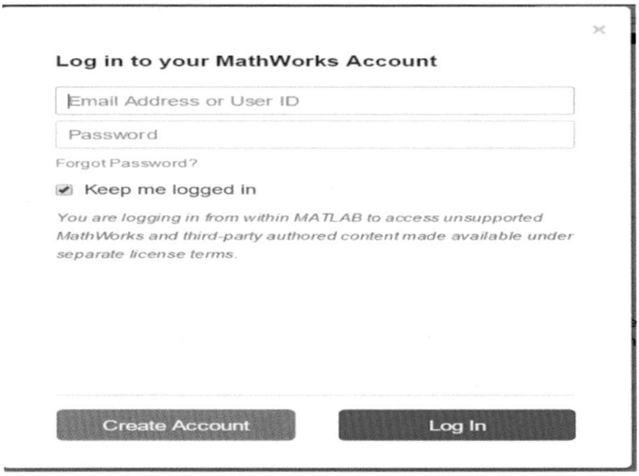

FIGURE 9.6   MathWorks account login.

4. You can make sure the support package has installed properly by typing arduinolib in the MATLAB command window. It will open up the library model with various blocks as shown in Fig. 9.9.

9.5   HARDWARE SETUP FOR THE DC MOTOR CONTROL

All the hardware parts described in Section 9.2 will be used here to set up the hardware and make the connections between the parts. Figs. 9.10 and 9.11 show the block diagram and electrical connections. The speed of the motor is controlled using the digital Pulse Width Modulated (PWM) output

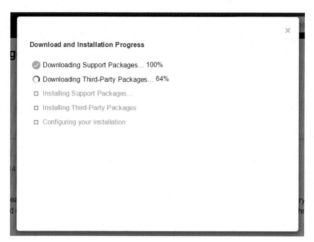

FIGURE 9.7 Support package installation progress.

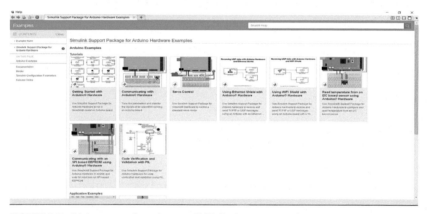

FIGURE 9.8 Link to example projects available in the support package.

of the Arduino Mega. The PWM output is used to switch the MOSFET which acts as a switch to connect and disconnect the 12V DC power to the DC motor. Although the motor is turned on and off continuously using the PWM output from the Arduino Mega, because of its inertia and friction it doesn't go to full speed when the PWM duty cycle command is 100%, or go to stop immediately when the PWM duty command goes to 0%. The motor dynamics (inertia and friction) act as a filter for the high frequency on-off commands given to the motor, and a smooth continuous motor speed profile can be obtained. The Arduino library provides a PWM output generator block to generate the PWM with a fixed period and varying duty cycle on a selected PWM output pin on the Arduino Mega board.

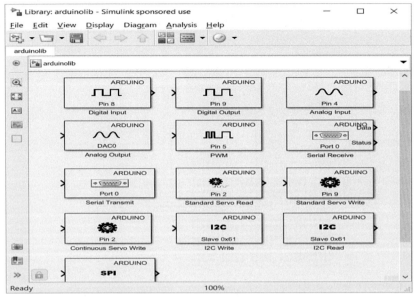

FIGURE 9.9 Arduino I/O library.

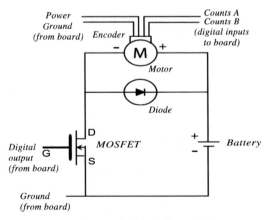

FIGURE 9.10 Hardware setup and connection block diagram [1].

The built-in quadrature encoder of the DC motor will be used in this experiment to measure the speed of the motor. The Arduino Support Package library does not provide a block to read the encoder output from the motor, so a custom S-function block is used to read the speed sensor.

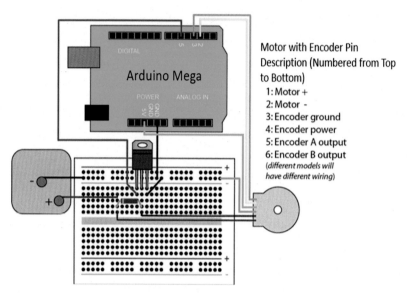

Motor with Encoder Pin
Description (Numbered from Top
to Bottom)
1: Motor +
2: Motor -
3: Encoder ground
4: Encoder power
5: Encoder A output
6: Encoder B output
(different models will
have different wiring)

FIGURE 9.11   Connecting Arduino to motor [1].

9.6   DATA COLLECTION FOR RESPONSE CURVE GENERATION

To assess nonlinearity of the plant, we will collect input/output data by sweeping the PWM signal across its full operating range and recording the resultant motor speed. This will allow us to plot the response curve of input versus output (steady state response). The DC motor is a single-input single-output system as shown in Fig. 9.12. The PWM duty cycle command is the input and the motor speed is the output. The input to Arduino PWM block is a value between 0 and 255. The block generates a PWM duty cycle between 0% and 100% on the configured hardware pin. The frequency of the waveform is approximately constant at 490 Hz.

We will sweep the PWM command from 0.25 to 255 with increments of 25 (except for the last step which ends in 255). The command will be changed every 50 seconds and the corresponding motor speed output is logged. Fig. 9.13 shows the PWM input waveform generated for response curve generation.

Follow the steps below to create and configure a Simulink model to run on Arduino Mega to sweep the PWM input and record corresponding motor speed output.

1. Open a new Simulink model and save it with name you want to use for the model.
2. Go to the *Model >> Simulation >> Model Configuration Parameters >> Solver* setting page and make the highlighted changes as shown in

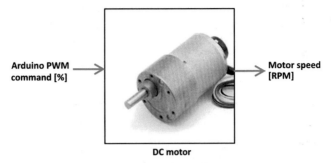

FIGURE 9.12 Input/output of a DC motor speed control system.

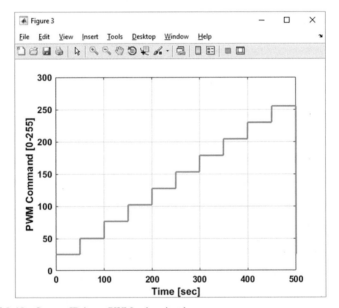

FIGURE 9.13 System ID input PWM value signal.

Fig. 9.14. Simulation *Stop Time* of 500 seconds is selected because as per Fig. 9.13—we need 500 seconds to sweep the input changes. *Solver Type* is changed to *Fixed-step* because all the blocks and subsystems in this model are expected to run at a certain discrete rate of 0.02 seconds specified in the *Fixed-step size* field.

3. Next, go to the Hardware Implementation configuration setting and select *Arduino Mega 2560* as the *Hardware board* as shown in Fig. 9.15. After making these changes, click *Apply* and *Ok* on the configuration setting GUI.

4. Add a *Repeating Sequence* block from the Simulink library to the newly created model, double click on the *Repeating Sequence* block and enter

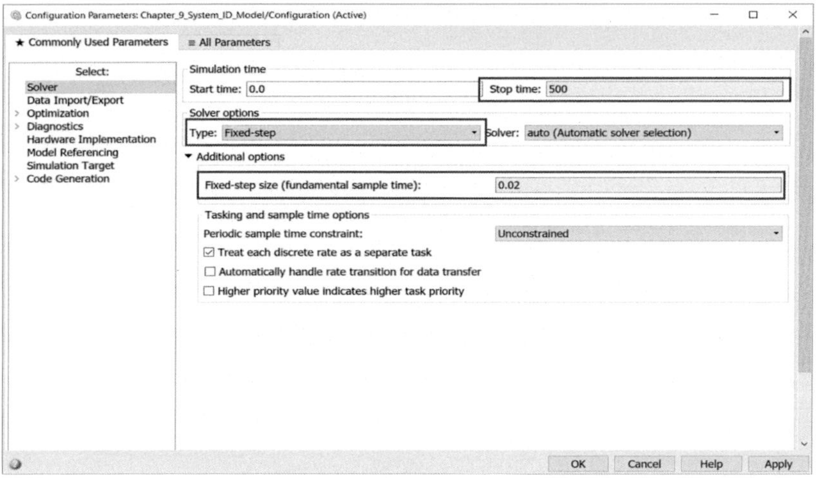

FIGURE 9.14 System ID model solver settings.

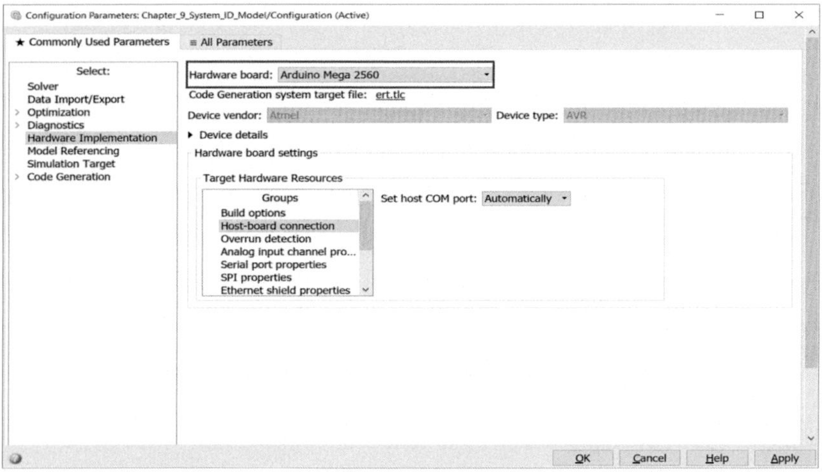

FIGURE 9.15 Hardware board selection.

the *Time values* and *Output values* to match the input sweep waveform shown in Fig. 9.16.

Time Values: [0 49.9 50 99.9 100 149.9 150 199.9 200 249.9 250 299.9 300 349.9 350 399.9 400 449.9 450 500] *Output Values*: [25.5 25.5 50.1 50.1 76.5 76.5 102 102 127.5 127.5 153 153 178.5 178.5 204 204 229.5 229.5 255 255] (Fig. 9.17).

5. Type *arduinolib* on the MATLAB command window to open up the Arduino Hardware support library. From the library, add the *PWM* block into the model and connect the output of the *Repeating Sequence*

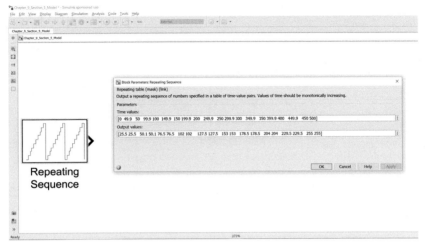

FIGURE 9.16 System ID input values in repeating sequence block.

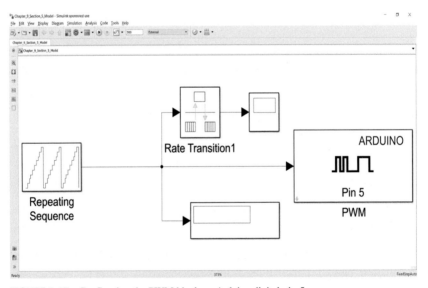

FIGURE 9.17 Configuring the PWM block on Arduino digital pin 5.

block to the input of the *PWM* block. Also, double click on the *PWM* block and enter the *Pin number* as 5 (if not already there). This is to configure the Digital pin 5 of the Arduino to drive the PWM output.

6. Just to make sure this Simulink model can communicate with the hardware setup made in Section 9.4, we can try to build and download this simple model into the hardware. Make sure all the connections in

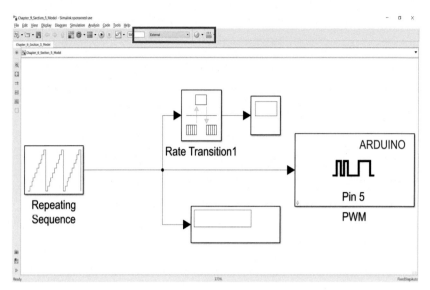

FIGURE 9.18 Selecting external mode simulation on the model.

Section 9.4 are proper. Now connect the Arduino Mega to computer running MATLAB using the USB cable which comes with the Arduino board. Change the *Simulation mode* of the model to *External*, as shown in Fig. 9.18. The external mode lets us tune the parameters in, and monitor data from, the Simulink model while it is running on the Arduino hardware. Also connect a scope and a display block to the output of the *Repeating Sequence* block through a rate transition block to monitor the value while it is running on the Arduino board.

7. Click on the simulation button on the model. Now Simulink will generate a code from this model, compile it and download the executable into the Arduino Mega hardware. If all the connections are proper after the download is completed, you will observe the motor starting to run at an initial PWM command of 25.5, and every 50 seconds the motor speed will increase. If you get any errors during the simulation build process, it is most likely that the Arduino board is not properly connected to the computer. If the simulation successfully starts but the motor is not spinning, it is probably because the connections are not proper as recommended in Section 9.4.

8. Proceed further if Step 7 is successful and the motor is running with an increasing speed after every 50 seconds until 500 seconds.

9. Now the Simulink logic required to sense the speed of the motor needs to be added. As mentioned earlier in this chapter the Arduino Support Package does not provide an in-built block in the library to read the encoder output. So a custom S-function obtained from the

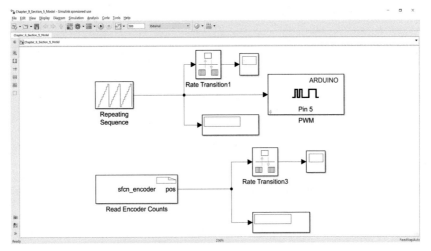

FIGURE 9.19 Adding the encoder read S-function.

guide [2] is used for the encoder position reading. For more details about the s-function please refer to the guide [2]. Add the encoder block from the package downloaded from the above reference link (or copy it from the *arduino_encoder_lib.slx* available in the attached folder, which is essentially the same s-function, but copied into a new library for the use of this chapter). Add scope and display blocks to the output of the encoder block through a rate transition block and click on the *Simulation* button again (Fig. 9.19).

10. Now when the simulation runs, the scope and display blocks connected to the output of the encoder block will show the encoder count readings as shown in Fig. 9.20. However, it can be seen that the encoder counts starts going up from zero, but after about 13 seconds it reaches a certain value and rolls over. This is because the S-function uses int16 as the datatype for the output, int16 can only hold values between 2^{15} (-32768) and 2^{15} (32768)—15 bits for the magnitude and one bit is used to hold the sign of the value.

11. The speed of the motor will be calculated from the encoder counts. The encoder output indicates the motor's position. The motor speed can be approximated over a specific time interval as the change in motor position between samples divided by the change in time. This will give the average speed of the motor over the time interval. The highlighted logic in Fig. 9.21 first takes a difference between the encoder positions between two samples using the *Unit Delay* block and divides the output by the sample time of the model which is 0.02 seconds to give the encoder counts/seconds. The counts/seconds needs to be converted to revolutions/second using the gear ratio of the motor being used. The

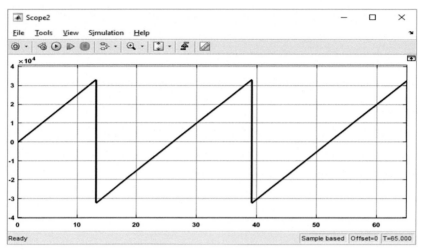

FIGURE 9.20 Output of encoder block.

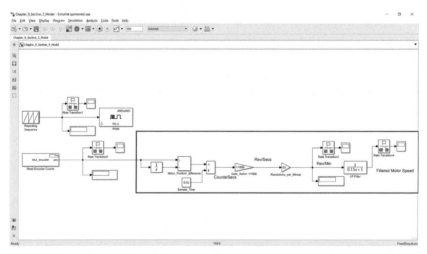

FIGURE 9.21 Motor speed calculation from encoder counts.

gear ratio of the motor used for this application is 1/1856 which corresponds to 1856 counts per revolution of the gearbox's output shaft. So, in order to convert the counts/seconds to revolutions/seconds the counts are divided by 1856 to give the motor speed in revolutions/seconds. Then it is multiplied by 60 to convert the revolutions/seconds to revolutions/minute. After adding the logic hit the *Simulation* again to run the updated logic on the hardware. Also add a scope block to monitor and

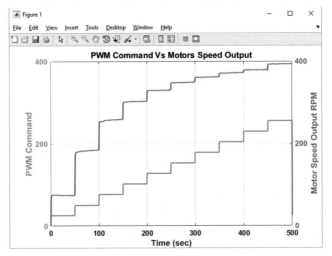

FIGURE 9.22   PWM input command value versus motor speed RPM.

log the motor speed output. Fig. 9.22 shows the PWM commanded by the Arduino megaboard and the corresponding motor speed RPM output.

12. A closer look at the sensed motor speed signal, even for the fixed steady PWM input command, shows there is some variation in the speed. This is attributed to noise in the sensing mechanism. It is recommended to use a filter to smooth out any noise in the measurements before the sensed value is used for controls purpose. A filtering was not really required in the previous chapters while the data for system identification is collected because only simulation models were used in those chapters and the measurement noise was not really modeled there. A simple first-order filter logic using the Simulink *Transfer Function* block is added to filter the motor speed, as shown in Fig. 9.23. A filter will cause a lag in the filtered value. The filter constant needs to be carefully chosen so that there is a balance between the output to input signal lag and reduction of noise. A filter constant of 0.15 seconds is used in this example. Fig. 9.24 shows the response of the filter.

9.7   ANALYZING SYSTEM NONLINEARITY

In this section, the data collected from Section 9.5 is analyzed to assess the nonlinearity of the DC motor. This analysis will enable us to determine how many MPC controllers will be needed. The steady state points of Fig. 9.22 been captured in Table 9.2. These points have been plotted in Fig. 9.25.

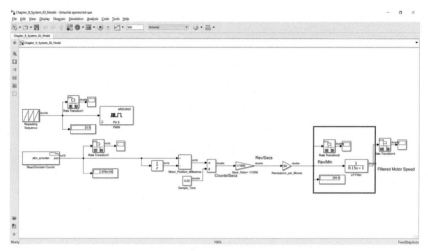

FIGURE 9.23 Filtering motor speed.

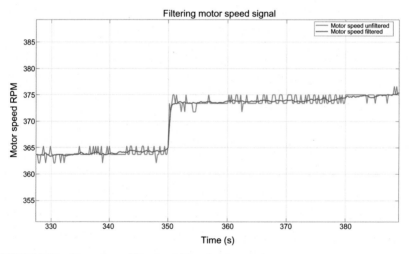

FIGURE 9.24 Comparing unfiltered and filtered motor speed.

Based on Fig. 9.25, we observe three regions of slope changes—marked as red, green and blue vertical lines. The system identification will be done using the data in these three regions separately. These models will then be used to design three MPC controllers. Run *Chapter_9_Section_7_Script.m* to replot Fig. 9.25 (Table 9.3).

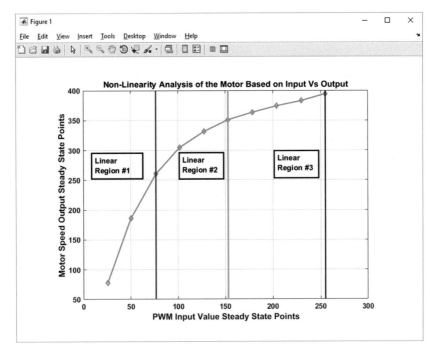

FIGURE 9.25 Nonlinearity assessment of the motor system.

Chapter_9_Section_7_Script.m

```
%Book Title: Practical Design and Application of MPC
%Chapter: 9
%Section: 7
%Authors: Nassim Khaled and Bibin Pattel
%Last Modified: 10/07/2017
%%
close all
fig=figure;
hax=axes;
hold on
input_pwm_steady_state_points = [25.5  50.1     76.5    102     127.5   153     178.5
204     229.5   255];
output_motor_speed_points = [78.1  185.9       260.2   305.1   331.5   351     363.3
374.1   382.7   394.4];
plot(input_pwm_steady_state_points,output_motor_speed_points,'linewidth',2,'Marker','diamo
nd');
line([76.5 76.5],get(hax,'YLim'),'Color',[1 0 0],'linewidth',2)
line([153 153],get(hax,'YLim'),'Color',[0 1 0],'linewidth',2)
line([255 255],get(hax,'YLim'),'Color',[0 0 1],'linewidth',2)
grid on
xlabel('PWM Input Value Steady State Points','FontSize',18);
ylabel('Motor Speed Output Steady State Points','FontSize',18)
set(gcf,'color',[1 1 1]);
title('Non-Linearity Analysis of the Motor Based on Input Vs Output','FontSize',18)
annotation(fig,'textbox',...
    [0.1635 0.577991452991452 0.0948333333333331 0.0512820512820514],...
'String',{'Linear Region #1'},...
'LineWidth',2,...
'FitBoxToText','off');
annotation(fig,'textbox',...
    [0.389802083333333 0.579594017094016 0.0948333333333332 0.0512820512820514],...
'String',{'Linear Region #2'},...
'LineWidth',2,...
'FitBoxToText','off');
annotation(fig,'textbox',...
    [0.649697916666666 0.582799145299144 0.0948333333333332 0.0512820512820514],...
'String',{'Linear Region #3'},...
'LineWidth',2,...
'FitBoxToText','off');
```

TABLE 9.3 Steady State Operating Points

PWM Command SS Value	25.5	50.1	76.5	102	127.5	153	178.5	204	229.5	255
Motor Speed SS Value	78.1	185.9	260.2	305.1	331.5	351	363.3	374.1	382.7	394.4

TABLE 9.4 PWM Step Values Used for System Identification

Region	Nominal PWM Value	PWM Step Final Value
Linear Region #1	50.1	76.5
Linear Region #2	127.5	153
Linear Region #3	204	229.5

9.8   SYSTEM IDENTIFICATION

The time series input/output data from the Fig. 9.22 will be used in this section for system identification. Filtered speed, as described in step 12 of Section 9.7, will be used to reduce the effect of measurement system noise.

We have the option to choose region #1 full data set (which includes two step responses) or to use a portion of the data. We chose the 1 step response in each region to do system identification. Assuming linearity in each region, choosing which portion of the data to use won't considerably affect the MPC controller.

Table 9.4 shows the range of data that was used in the system identification.

Chapter_9_Section_8_Script.m is used to extract the data shown in Table 9.4, perform system identification, plot the identified model against the measured data, and save the identified linear models into a .mat file. The order of linear models used in the three regions is two. Figs. 9.26 through to 9.28 shows the results of the system identification. The input (PWM command) and output (motor speed) have been offset to start from zero—the offset values regarding the nominal values as shown in Table 9.4. For more details about the system identification script and the MATLAB functions used, refer to Chapter 3, MPC Design of a Double-Mass Spring System and Chapter 4, System Identification for a Ship. The script *Chapter_9_Section_8_Script.m* can be found in Chapter_9/Section_8 folder.

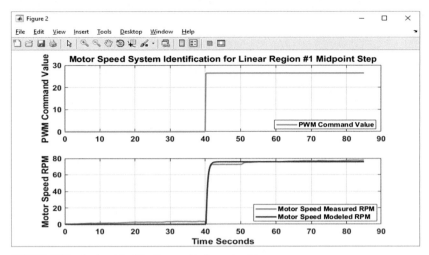

FIGURE 9.26 System identification results for region #1.

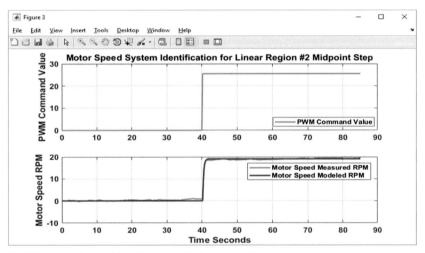

FIGURE 9.27 System identification results for region #2.

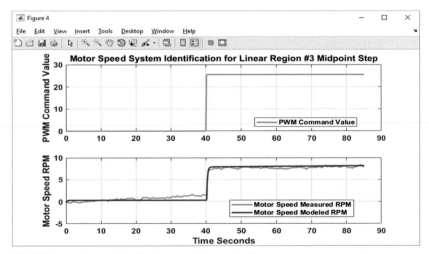

FIGURE 9.28 System identification results for region #3.

Chapter_9_Section_8_Script.m

```
%Book Title: Practical Design and Application of MPC
%Chapter: 9
%Section: 8
%Authors: Nassim Khaled and Bibin Pattel
%Last Modified: 10/07/2017
%%
clc
clear all
% Specify the path to look for the simulation results
curr_working_folder = pwd;
cd('..');
system_id_data_path = [pwd '\Section_5'];
cd(curr_working_folder);
% Load the system id data from Section 9.5
load_str = ['load ' system_id_data_path '\Chapter_9_Section_6_System_ID_Data.mat '];
eval(load_str);


%% Model # 1 Grouping simulation data and neglecting first few seconds(Nstart*stp_sz)
% Specify sample time of data
stp_sz = 0.2;
Nstart = 300;%50; % Clip the data starting from time t = 10 Secs to t = 145 Secs
Nend = 725;%
Measured_Outputs= Motor_Speed_RPM_Filtered(Nstart:Nend,2); % Measured outputs
Manipipulated_Variables= PWM_Input(Nstart:Nend,2); %Manipulated variables

% Obtaining initial conditions at step time=Nstart
Measured_Outputs_Nstart=Measured_Outputs(1,:); %Capturing the measured outputs at step
time= Nstart
Manipipulated_Variables_Nstart=Manipipulated_Variables(1,:); %Capturing the manipulated
variables at step time= Nstart

% Forcing response to start from zero initial conditions
Measured_Outputs_zero_initial_conditions=Measured_Outputs-
repmat(Measured_Outputs_Nstart,length(Measured_Outputs),1); %Subtracting initial
conditions for measured outputs to obtain zero response at step time= Nstart
Manipipulated_Variables_zero_initial_conditions=Manipipulated_Variables-
repmat(Manipipulated_Variables_Nstart,length(Manipipulated_Variables),1); %Subtracting
initial conditions for manipulated variables to obtain zero actuation at step time= Nstart

% Prepare date for system identification
data=iddata(Measured_Outputs_zero_initial_conditions,Manipipulated_Variables_zero_initial_
conditions,stp_sz); %data is packaged for system identification using idda-ta

% Generate a preliminary 2nd order system that fits the data
sys1=n4sid(data,2,'Form','canonical','DisturbanceModel','None','InputDelay',[0
0]','InitialState','zero'); %n4sid generates a preliminary system in the canonical form
%with zero disturbance, zero delay and zero initial conditions

% Generate a more refined system
sys2 = pem(data,sys1,'InitialState','zero') %pem generates a more refined system that fits
the data better

% Define the options for comparing the various identified systems
opt = compareOptions('InitialCondition','Z');
[Y,fit,x0]=compare(data,sys2);
Y_1=Y.OutputData(:,1);

figure
subplot(2,1,1)
plot(0:stp_sz:stp_sz*(length(Y_1)-
1),Manipipulated_Variables_zero_initial_conditions(:,1),'linewidth',2)
grid on
legend('PWM Command Value')
ylabel('PWM Command Value','FontSize',18)
title('Motor Speed System Identification for Linear Region #1 Midpoint
Step','FontSize',18);
subplot(2,1,2)
plot(0:stp_sz:stp_sz*(length(Y_1)-
1),Measured_Outputs_zero_initial_conditions(:,1),'linewidth',2)
grid on
hold on
plot(0:stp_sz:stp_sz*(length(Y_1)-1),Y_1,'r','linewidth',2)
legend('Motor Speed Measured RPM','Motor Speed Modeled RPM')
```

```
ylabel('Motor Speed RPM','FontSize',18)
xlabel('Time Seconds','FontSize',18)
set(gcf,'color',[1 1 1]);

eval_str = 'Motor_Linear_Model_Mode_1 = sys2;';
eval(eval_str);
eval_str = ['save Motor_Linear_Model_Mode_1.mat'' Motor_Linear_Model_Mode_1'];
eval(eval_str);
```

9.9   MPC CONTROLLER DESIGN

In this section, the linear models identified will be used to design the MPC controllers. The steps of designing the controller using the *MPC Designer App* for the linear region #1 is described here. The same steps can be followed for region #2 and #3.

1. Copy the three linear models developed in Section 9.8 into a new folder named Section_8 and load the linear model for region #1 into the MATLAB workspace using the command *load ('Motor_Linear_Model_Mode_1.mat')*.

2. Open the MPC Designer GUI by typing *mpcDesigner* in the MATLAB command window.

3. The GUI has three submenu lists under *MPC Design Task* Menu. The *Plant Models* menu is where you can import plant models, the *Controllers* menu allows you to select the imported plant model/models to develop the MPC controller and specify the constraints and allows weight tuning, among others. The *Scenarios* menu is where you can test the designed controller with the plant model to make sure the controller is working as expected. We can iterate by going back to the *Controllers* menu and redesign and test again if the performance is not acceptable.

4. Click on the Button *MPC Structure* on the GUI. This will pop up another GUI, as shown in Fig. 9.29. When this GUI is opened it will search in the workspace for any available state space models and display it for the User to select to import. After that click on the row to select the plant model you want to import to the MPC Design GUI. Select *Motor_Linear_Model_Mode_1*. It will show the properties of the selected plant. Click the *Define and Import* button and the designer will create a default MPC controller structure with the imported plant, controller, and scenario as shown in Fig. 9.30.

5. Under the Controllers menu there will be a default controller called mpc1 added automatically by the *Define MPC Structure* GUI. Users can change the name of this controller—click on the name of the controller three times repeatedly and it will display the name in an edit box to edit the name. Also, it will add a scenario1 as shown, which can be renamed as well. In this example, we renamed the controller to be *Motor_Linear_MPC_Mode_1* and the scenario1 to be *Motor_Linear_MPC_Mode_1_Test_1* (Fig. 9.30).

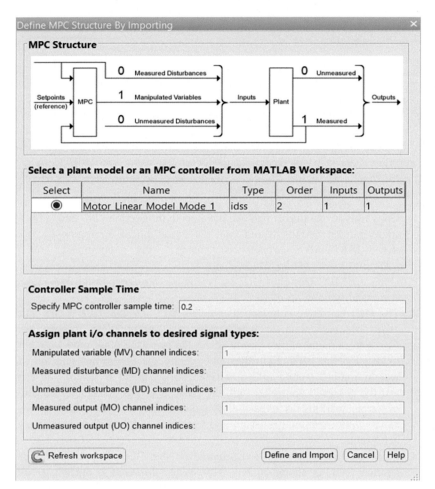

FIGURE 9.29   Plant model importer GUI.

6. Click on the *I/O Attributes* button on the GUI to enter the nominal values of input and output for the plant as shown in Fig. 9.31. The nominal values are based on Table 9.4.
7. Click on the *Tuning* tab at the top of the *MPC Designer GUI*. The controller design window will be displayed at the top of this window. Here, you can select the controller and plant from the drop-down menu. The drop-down menu to select a plant model can be used if you have multiple plant models to select from. Select the *Motor_Linear_Model_Mode_1* which was imported earlier. After selecting the plant, we need to specify the sample time of the controller. The GUI automatically takes the sample time of the imported plant as the *Sample Time* (in seconds) as shown, in this case 0.2 seconds. This can

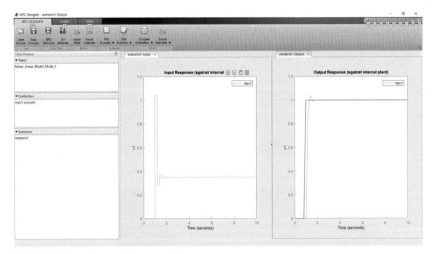

FIGURE 9.30   Plant model imported.

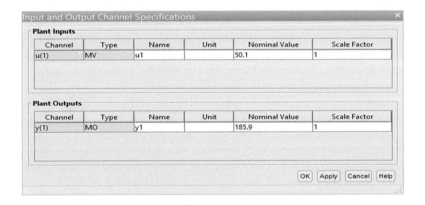

FIGURE 9.31   Specify plant input and output nominal values.

be changed to a different value if you want to run the controller at a different rate than that of the plant model. Additional rate transition blocks might be needed to handle it when you run the simulation in Simulink. In this case, you will use the same control interval as that of the plant. Next are the prediction and control horizon intervals. By default, the GUI will use a prediction horizon of 10 intervals which is 2 seconds and a control horizon of 2 intervals which is 0.4 seconds, according to the control interval (controller sample time) having been selected as 0.2 seconds. The higher the prediction and control horizon, more computational and memory resources it will require. The

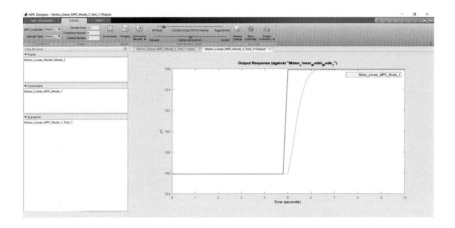

FIGURE 9.32 Selecting the model and controller horizons.

prediction is changed to 8 samples and the control horizon is set as default 2. These settings can be revisited later if the controller does not meet the desired performance. See Fig. 9.32.

8. Click on the *Constraints* button on the ribbon to specify the input and output constraints the MPC controller design needs to meet. The constraints can be specified for input and output variables separately here. For the input manipulated variables, the minimum and maximum allowed values and the rate of increment and decrement for the control action can be constrained by entering values into the cells provided. In this example, we are constraining the input command to range from 0 to 255, and the rise and fall rate of the control action to be the default −inf and inf. Similarly, the output motor speed is constrained between 0 to 400 RPM (Fig. 9.33).

9. Click on the *Weights* button on the ribbon to specify the input and output weights. This section allows you to adjust the controller performance by specifying weights on the input manipulated variables, the outputs, or both. The *Input Weights* table lists all the manipulated variables of the controller. For each MV, we can specify the *Weight* and *Rate Weight* options. The *Weight* option penalizes the change in the manipulated variable from its nominal value specified (see Fig. 9.34). We can keep this cell either 0 or blank so as not to penalize the control action. This means the manipulated variable can move anywhere between its lower and upper bounds provided in the *Constraints* option. Or we can give a positive value for the *Weight* to penalize the manipulated variable to keep it closer to its nominal value. In this case, we kept the Weight as zero for the manipulated variable. The *Rate Weight*

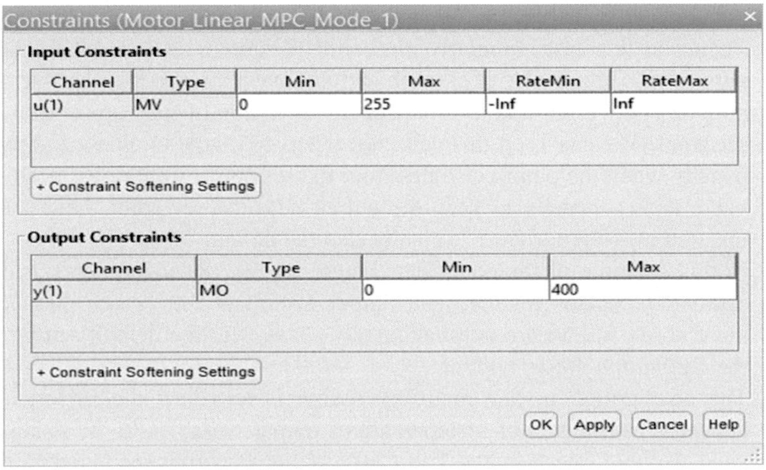

Constraints (Motor_Linear_MPC_Mode_1)

Input Constraints

Channel	Type	Min	Max	RateMin	RateMax
u(1)	MV	0	255	-Inf	Inf

+ Constraint Softening Settings

Output Constraints

Channel	Type	Min	Max
y(1)	MO	0	400

+ Constraint Softening Settings

OK   Apply   Cancel   Help

FIGURE 9.33 Specifying the constraints for MPC.

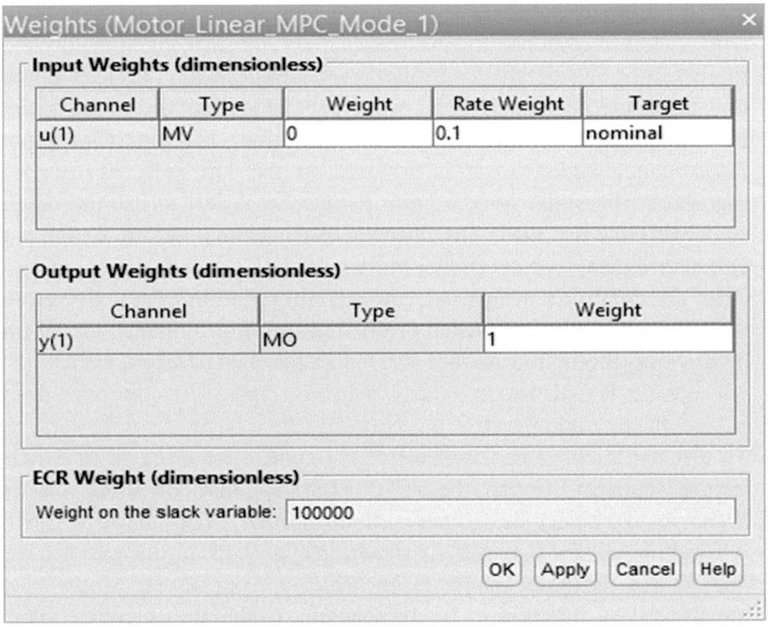

Weights (Motor_Linear_MPC_Mode_1)

Input Weights (dimensionless)

Channel	Type	Weight	Rate Weight	Target
u(1)	MV	0	0.1	nominal

Output Weights (dimensionless)

Channel	Type	Weight
y(1)	MO	1

ECR Weight (dimensionless)

Weight on the slack variable: 100000

OK   Apply   Cancel   Help

FIGURE 9.34 Specifying controller weight tuning.

option penalizes the change in value of a manipulated variable. We can keep this cell either 0 or blank so as not to penalize the rate change, the manipulated variable can move up or down at any rate within the bounds provided in the *Constraints* option. A positive value for the *Rate Wight* applies a penalty to reduce the rate of change of the manipulated variable. Let us keep the default value as 0.1 for the *Rate Weight* in this case. Similarly, the *Input Weights* table lists all the outputs of the controller. For each output, we can specify the *Weight* option. The *Weight* option penalizes the deviation of the output from its set point. We can keep this cell either zero or blank so as not to apply penalty when the output deviates from the set point. Applying a positive value adds a penalty to keep the output near the set point. Let's keep this output weight to be 1, which is also the default.

10. In this example of the controller design, we are not using any specific estimation models for the plant output disturbance or sensor measurement noise. So, we are not making any changes to the default setting of the *Estimation Models* option.

11. The next option in the controller design is to adjust the importance between the weights of manipulated or output variables to the manipulated variable rate weights. The *Closed Loop Performance* slider in the GUI adjusts the weights on all the input and output variables. Moving the slider to the left increases manipulated variable rate penalties relative to setpoint penalties, often increasing controller robustness, but with a slower response to disturbances. By contrast, moving the slider to the right makes the controller more aggressive, but less tolerant towards modeling errors. Some overshoots were observed in the controller performance with the default position of the *Closed-Loop Performance* slider, so it is adjusted to the left side, as shown in Fig. 9.32. The slider for the *State Estimation* is left to use the default, since we have not made any changes to the default setting of the plant output and measurement model in Step 10.

12. After the controller design, we can perform a design review to examine the MPC controller for design errors and stability problems at run time by clicking the option *Review Design* in the *MPC Designer* GUI. This will open a HTML report as shown in Fig. 9.35. The controller design passed all the review tests.

13. To test the tracking performance of the controller, we need to define a testing scenario. Go to the *MPC Designer* tab and click on *Edit Scenario* and select the scenario *Motor_Linear_MPC_Mode_1_Test_1*. A window will be displayed as shown in Fig. 9.36. In the scenario window, select the plant model to be *Motor_Linear_Model_Mode_1*, and the simulation duration to be 10 seconds. Under the reference signals section under the *Signal* column, select *Step* input for the output reference. With a default nominal value of 185.9 for the step input, change the step size to be 10 for the reference, but change the time at which the step happens for y_1 to be after 5 seconds. Let's not add any output or load disturbances in this example. Click on the *Apply* and the *Ok* buttons. The MPC Designer tool will now simulate the selected plant model with the MPC controller designed in the above steps along with

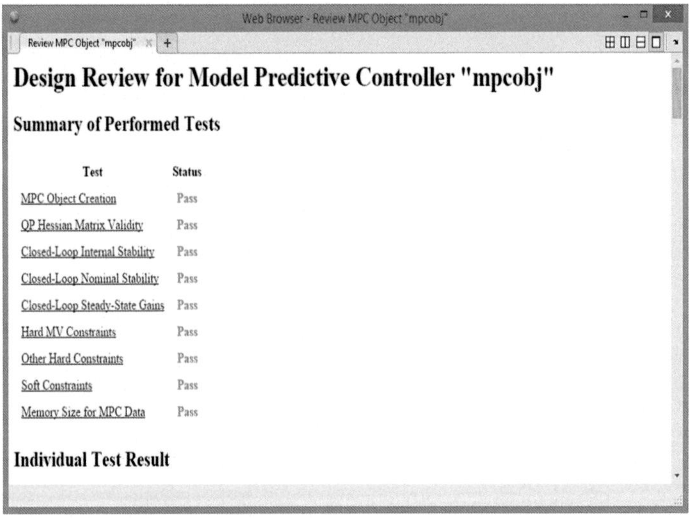

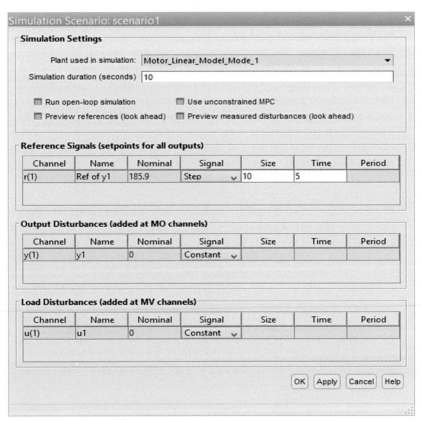

FIGURE 9.35 Design review of MPC.

FIGURE 9.36 Editing the test scenario to validate MPC controller.

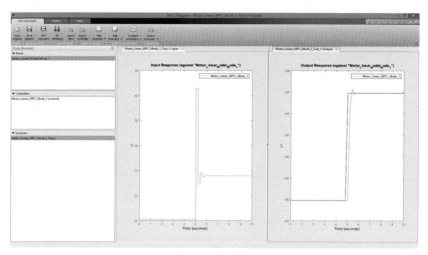

FIGURE 9.37 MPC controller performance tracking with the test scenario.

the reference tracking outputs defined in the scenario window and plot the MPC controller commands and the MPC controller performance tracking as shown in Fig. 9.37.

14. We can change the various settings in the *Tuning* tab such as *Constraints*, *Weights*, *Prediction*, and *Control Horizon*, etc., and see the controller performance with the same scenario instantly. After satisfactory validation of the controller, we can save the controller design project into a .mat file with all the settings in the *MPC Designer* GUI. This will help users to open the controller design at any point of time and make tweaks as desired. Use the *Save Session* button to save it into a *.mat* file. This example design is saved as *MPC_DesignTask_Chapter_9_Section_9_MPC_1.mat* and can be found in the *Chapter_9/Section_9* folder.

15. We can also export the MPC Controller into the MATLAB workspace using the *Export Controller* option. Fig. 9.38 shows the *Export Controller* GUI. Users will be able to change the name of the controller object name while exporting, if needed, by changing the name in the *Export As* column. After exporting the controller into the MATLAB workspace, we can type in the controller object name in the MATLAB command window to see the structure and properties of the controller object, as shown in Fig. 9.39. Save the MPC controller into a *.mat* file by typing the command *save Motor_Linear_MPC_Mode_1.mat Motor_Linear_MPC_Mode_1* in the MATLAB command window.

16. Repeat steps 1 to 15 for regions #2 and #3. Fig. 9.40 shows the resultant MPC controllers saved in *.mat* format.

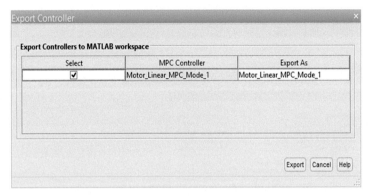

FIGURE 9.38 Exporting the MPC controller into the MATLAB workspace.

```
Command Window
New to MATLAB? See resources for Getting Started.

>> Motor_Linear_MPC_Mode_1

MPC object (created on 12-Oct-2017 09:31:31):
---------------------------------------------
Sampling time:     0.2 (seconds)
Prediction Horizon: 10
Control Horizon:   2

Plant Model:
                          --------------
    1 manipulated variable(s)   -->|  2 states  |
                              |             |--> 1 measured output(s)
    0 measured disturbance(s)   -->|  1 inputs  |
                              |             |--> 0 unmeasured output(s)
    0 unmeasured disturbance(s) -->|  1 outputs |
                          --------------
Disturbance and Noise Models:
        Output disturbance model: default (type "getoutdist(Motor_Linear_MPC_Mode_1)" for details)
        Measurement noise model: default (unity gain after scaling)

Weights:
        ManipulatedVariables: 0
    ManipulatedVariablesRate: 0.1000
            OutputVariables: 1
                    ECR: 100000

State Estimation:  Default Kalman Filter (type "getEstimator(Motor_Linear_MPC_Mode_1)" for details)

Constraints:
    0 <= u1 <= 255, u1/rate is unconstrained, 0 <= y1 <= 400
>> save Motor_Linear_MPC_Mode_1.mat Motor_Linear_MPC_Mode_1
```

FIGURE 9.39 Displaying the MPC structure and main parameters.

9.10   INTEGRATING MPC CONTROLLERS WITH SIMULINK MODEL

In this section, the MPC controllers designed in Section 9.9 will be integrated with Simulink. It is advised that the reader opens *Chapter_9_Section_10_Model_1.slx* model while reading through this section. A top-level view of the simulation model is shown in Fig. 9.41:

The main components of the model are:

- Motor speed reference system
- Motor speed feedback system

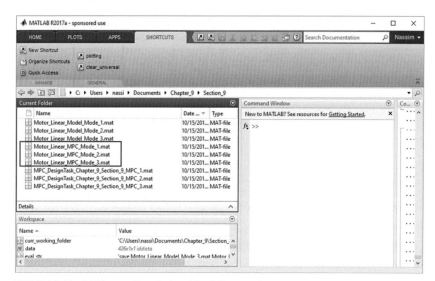

FIGURE 9.40 MPC controllers designed for 3 linear regions.

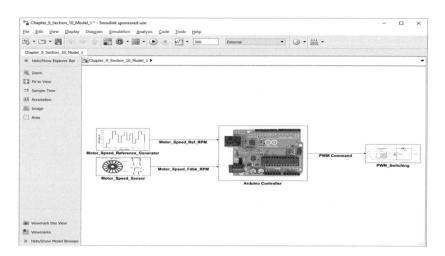

FIGURE 9.41 Top-level Simulink model of real-time motor speed controller using multimode MPC.

- MPC mode selection and control system
- PWM actuation and switching system

The reference system generates a reference signal for the motor speed. The reference is designed to push the controller across multiple regions of the operating space. Fig. 9.42 shows a time series plot for the reference

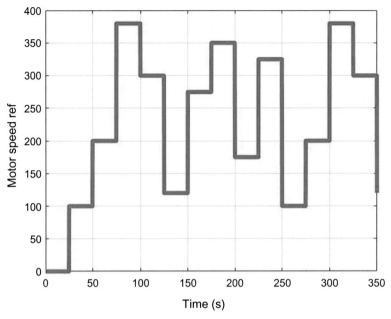

FIGURE 9.42 Motor speed reference for the MPC controller.

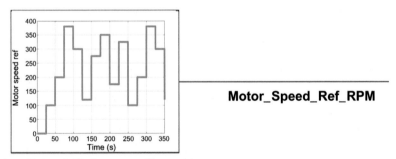

Motor_Speed_Reference_Generator

FIGURE 9.43 Reference generator subsystem, top view.

signal. Figs. 9.43 and 9.44 show the Simulink subsystem for the reference generation.

Figs. 9.45 and 9.46 show the Simulink logic for the motor speed sensing system. The logic developed in Section 9.6 for the motor speed sensing for system identification data collection is used here as well.

The MPC controller top level subsystem is shown in Fig. 9.47. This block takes the motor speed reference and motor speed feedback as inputs and outputs for the PWM command value.

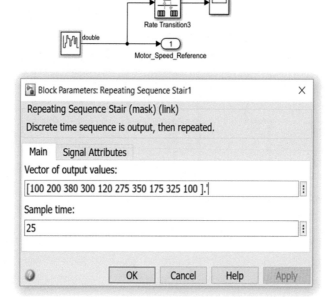

FIGURE 9.44   Reference generation using Simulink repeating sequence stair block.

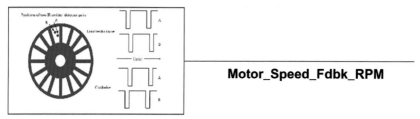

Motor_Speed_Sensor

FIGURE 9.45   Motor speed sensing subsystem top view.

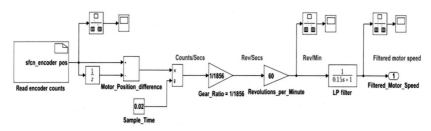

FIGURE 9.46   Motor speed sensing logic.

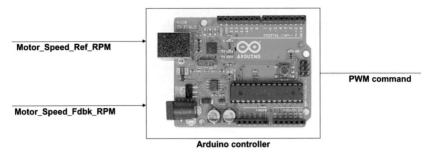

Motor_Speed_Ref_RPM

Motor_Speed_Fdbk_RPM

PWM command

Arduino controller

FIGURE 9.47 Motor controller subsystem top view.

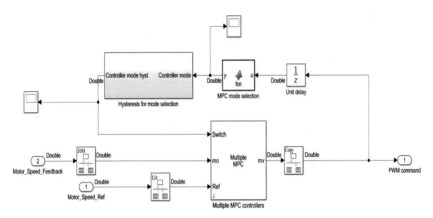

FIGURE 9.48 Multimode MPC controller logic.

A look inside the subsystem shows the hysteresis block, the mode selection logic, and multiple MPC block (Fig. 9.48). The PWM command is used to select which controller will be used among the three MPC controllers. The selection is based on where the PWM command falls in Table 9.4. The integer representing the region for the selected MPC controller is called a mode. The main components of multi-MPC controller are: Mode selection and the MPC controller.

MPC mode selection is implemented using a MATLAB function block that can be obtained from *Simulink/User-Defined Functions/MATLAB Function*. Users can double click on the block to open the m file editor and type in the logic. Fig. 9.48 shows the MPC selector block with the controller PWM command fed through a Unit Delay block to the mode selector MATLAB function block. It outputs an integer between 1 and 3. Fig. 9.49 shows the MATLAB function code that computes the mode based on Table 9.4. The selected mode output is passed through a Hysteresis logic

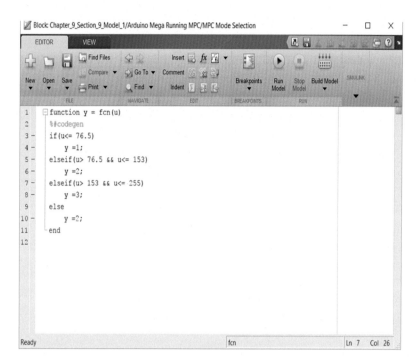

FIGURE 9.49 Function block for MPC mode selector logic.

(refer to Chapter 6, Multiple MPC Design for a Ship, for more details) to avoid rapid switching between the controller modes which can cause unwanted chattering in the controller performance. Hysteresis provides a smooth transition between different controllers.

The Simulink MPC library can be opened by typing *mpclib* on the MATLAB command window. Alternatively, it can be found in the Simulink library by searching for MPC. The MPC library shown in Fig. 9.50 with the *Multiple MPC* block is highlighted. Note that we unchecked the measured disturbance (md) option in the Multiple MPC block since we aren't using any measured disturbance in this controller design.

Fig. 9.51 shows the cell array of the MPC Controllers field and MPC Initial Controller States. As shown in Fig. 9.48, the multiple MPC block now has three inputs and one output. The first input is the MPC *switch* value from the mode selection logic, the second input is the *measured output (mo)* which is connected to the motor speed feedback signal and the third input is the *reference (ref)* which is the controller reference motor speed. The output of the multiple MPC controller block is the *manipulated variables (MV)* and the PWM command to achieve the desired motor speed.

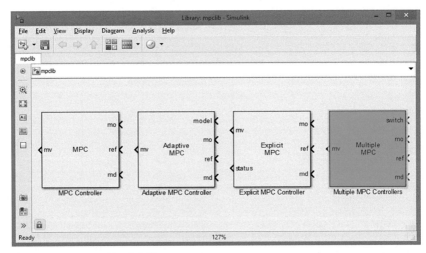

FIGURE 9.50   MPC Simulink library.

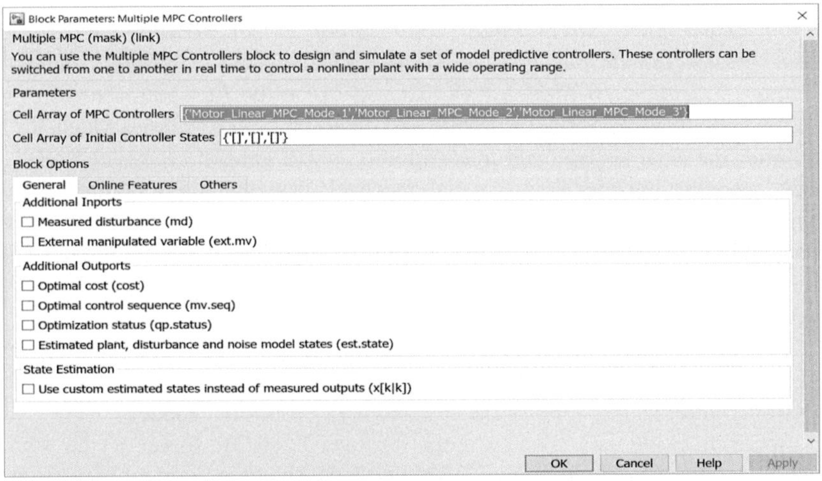

FIGURE 9.51   Settings for a multiple MPC block.

9.11 MULTIMODE MPC CONTROLLER DEPLOYMENT ON THE HARDWARE

The Simulink model from the previous section will be deployed into the Arduino Mega board to control the speed of the motor. Make sure the motor hardware and the Arduino Mega board are connected to the computer with an open session of MATLAB (preferably R2017a). Set the MATLAB directory to Chapter_9/Section_10 directory. In the MATLAB command

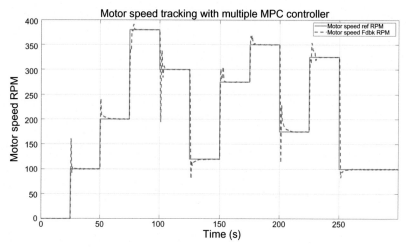

FIGURE 9.52   Real time motor speed tracking using the multiple MPC controller.

window, load the MPC controller .mat files by typing the commands: *load Motor_Linear_MPC_Mode_1.mat; load Motor_Linear_MPC_Mode_2.mat; load Motor_Linear_MPC_Mode_3.mat.* Open the Simulink model *Chapter_9_Section_10_Model_1.slx.* Then click on the *Run* button to generate the code from the model, compile and deploy into the Arduino Mega.

After the code is deployed into Arduino, it will generate the speed reference, read the encoder motor speed feedback, run the MPC controller to come up with the desired PWM command to track the motor speed, and send the PWM command to the PWM pin. All the signals can be monitored in real time using the scopes connected in the model. Signal logging can be enabled in the scopes to plot these signals later.

The performance of the controller motor speed tracking is plotted in Fig. 9.52. Fig. 9.53 shows the corresponding PWM command from the MPC controller and Fig. 9.54 shows the selected MPC controller mode.

Speed tracking is a good except for overshoots in some of the step responses of the tracking in Fig. 9.52. Fig. 9.55 shows a zoomed in version of Fig. 9.52. There are multiple parameters in the design to reduce overshoot. These will be proposed in the application problem for the reader to experiment with.

9.12   SINGLE MPC CONTROLLER DEPLOYMENT ON THE HARDWARE

To see how well a single MPC controller can control the nonlinear motor for a single mode MPC controller, the Simulink model is also created and deployed into the Arduino hardware while the speed tracking performance is

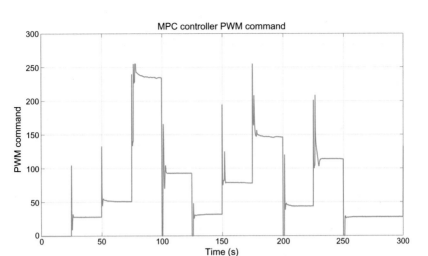

FIGURE 9.53 Multimode MPC controller output PWM command.

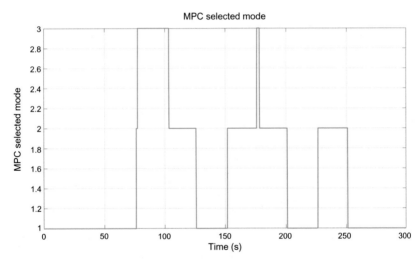

FIGURE 9.54 Selected modes for Multimode MPC controller.

logged. The Simulink model structure remains pretty much the same except that the MPC Controller subsystem no longer has the MPC selection or hysteresis logic, and the Multiple MPC block is replaced by a single MPC block. The MPC controller *Motor_Linear_MPC_Mode_2* is used in the MPC controller block as this is the controller designed for the middle region of the motor operating points. Fig. 9.56 shows the single mode controller logic in Simulink. Figs. 9.57 and 9.58 shows the performance of the single mode MPC controller. The single mode controller performance is better in regions

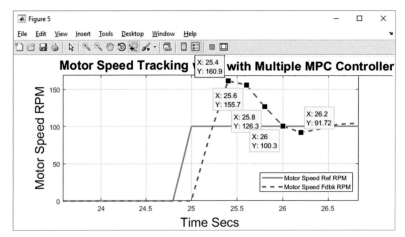

FIGURE 9.55 Zoom in for the real time motor speed tracking using multiple MPC controller.

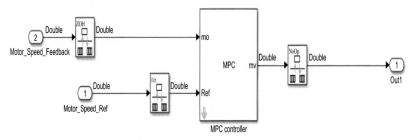

FIGURE 9.56 Single-mode MPC controller logic.

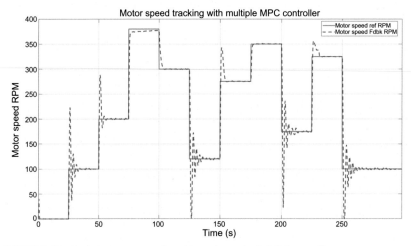

FIGURE 9.57 Real-time motor speed tracking using a single MPC controller.

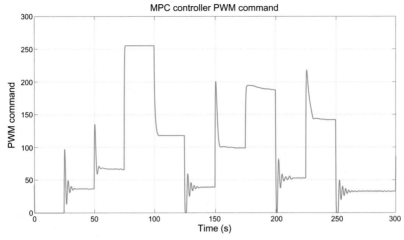

FIGURE 9.58   Single-mode MPC controller output PWM command.

where its model is identified (region #2). For other speed targets, eventually it ends up following the reference, but there are more overshoots and longer settling time.

9.13   APPLICATION PROBLEM

To solve the overshoot problem in Section 9.11, investigate the options below. Download *Chapter_9/Section_11* material and try to:

1. Add two more regions in the MPC design. Do you observer reduced overshoots?
2. Go back to the original design that has three MPC regions. The design has a sample time of 0.2 seconds. This means that the prediction horizon in the time domain is $0.2 \times 8 = 1.6$ seconds. Increase sampling time to 0.5 seconds and regenerate the results in Section 9.11. Do you observe reduced overshoots?
3. Go back to the original design that has three MPC regions and 0.2 seconds sample time. Change the aggressiveness of the controller (Fig. 9.32). Do you observe reduced overshoots?

REFERENCES

[1]   PI Control of a DC Motor. http://ctms.engin.umich.edu/CTMS/index.php?aux=Activities_ DCmotorB
[2]   Developing Simulink Device Driver Blocks: Step-By-Step Guide and Examples. http:// www.mathworks.com/matlabcentral/fileexchange/39354-device-drivers

Chapter 10

MPC Design for Air-Handling Control of a Diesel Engine

10.1 INTRODUCTION

In this chapter, we design and deploy a MPC controller for a diesel engine with dual exhaust gas recirculation (EGR) loops. This engine architecture requires the control of two EGR valves (high pressure and low pressure), an exhaust throttle (ET) and a variable geometry turbocharger (VGT). Following the same process reiterated throughout the book, and outlined in Fig. 10.1, we design and deploy the production-ready MPC. Unlike previous chapters where MATLAB and Simulink were used to design and simulate the controller, we use Honeywell's Automotive software, OnRamp Design Suite [1] and [2] in addition to MATLAB and Simulink.

In Section 10.2, we survey various air-handling control strategies for diesel engines. In Sections 10.3 and 10.4 we describe the architecture of the diesel engine and that of the air-handling system. The torque curve and the transient cycle are introduced in Section 10.5. System identification is implemented in Section 10.6. In Section 10.7, a MPC structure is described. In Section 10.8, we detail the process of deploying a MPC in the microcontroller. Testcell results and performance of the controller are discussed in Section 10.9. Furthermore, robustness analysis of the design is carried out in Section 10.10. A summary of this chapter is provided in Section 10.11. Related references are listed in Section 10.11.

The content of this chapter is reprinted with permission from SAE Copyright © 2014 SAE International. Further distribution of this material is not permitted without prior permission from SAE [3]. The authors would like to acknowledge the coauthors of the paper for their contributions [3].

10.2 AIR-HANDLING CONTROL SURVEY

Light duty vehicle emission standards are getting more stringent every year as stipulated by US EPA Tier 2 standards and LEV III regulations proposed by CARB. Dual loop EGR systems are known to have improved turbocharger operating efficiency, charge air temperature reduction, improved BSFC at steady state, and the capability to drive high rates of EGR while

Practical Design and Application of Model Predictive Control.
DOI: https://doi.org/10.1016/B978-0-12-813918-9.00010-1
© 2018 Elsevier Inc. All rights reserved.

FIGURE 10.1　Controller design and deployment framework.

FIGURE 10.2　ALTAS program.

minimizing the impact on performance [4] and [5]. Therefore, dual loop EGR systems offer significant advantages to reduce emissions and fuel consumption and can help meet strict emission requirements.

ATLAS is a Cummins program (Fig. 10.2) funded by the United States Department of Energy. The main objective of ATLAS is to demonstrate high fuel economy with significant emissions reduction for a small displacement engine (2.8 L) in a pick-up truck. The air-handling system is complex with a VGT, low pressure EGR valve, high pressure EGR valve, and an ET. Due to the higher complexity of the ATLAS engine, the coordination of the four actuators presents a challenging control and calibration hurdle for evaluating the transient performance of the engine.

Traditional control design techniques are typically based on standard PID controllers, lookup tables, and logical switches, etc., which have been around for many decades and are well known to both engineers and technicians. These techniques are relatively easy to use and implement, and may be ideal for control loops without interactions and with simpler dynamics. However, as the complexity of engines increase and the number of sensors and actuators grow, it becomes a difficult task to design a feedback controller using traditional decoupled loops within a reasonable timeframe. A more systematic approach using multivariable control techniques can help to reduce the development time while ensuring the required performance and emissions limits given by legislation.

Analysis of different EGR strategies can be found, e.g., in survey papers [6−8]. A cascaded adaptive PI strategy with model based feedforward was proposed in Shutty et al. [9]. A cascaded control structure of boost pressure served as a decoupling tool for EGR rate and boost pressure. In Grondin et al. [10] the strategy of dealing with HP and LP EGR separately by motion planning was introduced to control boost pressure and EGR rate. A cooperative control strategy for dual-loop EGR system was introduced in Yan et al. [11]. The authors used the fact that a LP-EGR loop has a substantially larger volume and consequently slower dynamics than those of the HP-EGR to decompose the original system into two separate subsystems with different time scales. For each subsystem, Lyapunov based controllers were designed and then collected together to control intake manifold pressure, temperature and oxygen fraction. In Shutty et al. [12] a coordinated strategy of dual-EGR, VGT, and intake throttle was shown with dynamic feedback of total EGR mass flow and boost pressure with HP/LP EGR split strategy to minimize pumping losses while maximizing the efficiency of the turbocharger.

Haber [13] designed a multivariable H∞ controller to manage the pressure, temperature, and oxygen concentration at the intake manifold. The controller commands the VGT position to mainly control the intake manifold pressure. Furthermore, the controller drives the LP and HP EGR valves to control the intake manifold temperature and oxygen concentration. A mean-value model for the air-handling system and air fraction estimator were developed in Wang [14].

Any multivariable approach mentioned above (i.e., H∞ and LQG) can be used to develop a systematic procedure for designing flexible and configurable controllers for air-handling systems. The main advantage of MPC is its ability to handle various constraints. A MPC can be an extension of LQG control with systematic handling of time-varying constraints. However, implementation of a MPC controller may be more complicated than simpler control algorithms because it requires running an optimization solver online at each sampling period. If the control problem is small enough (in terms of the number of total constraints in the resulting optimization problem) then the MPC control problem can be solved explicitly, where the resulting online solver is simple. It is based on a set of static lookup tables. More details about explicit MPC can be found in Bemporad et al. [15]. If the resulting problem cannot be solved explicitly due to a large number of constraints (e.g., unacceptable amount of memory to store all lookup tables) then it is still possible to implement some alternative and very efficient solver that can be implemented in the electronic control module (ECM) with reasonable CPU overhead, see, e.g., Borelli et al. [16].

10.3 ENGINE ARCHITECTURE

The engine configuration used in this study is described in Table 10.1. A Euro IV emissions compliant, inline 4 cylinders 2.8 L ISF engine (Fig. 10.3) was modified as described in detail in Suresh et al. [4]. The after-treatment system of the ATLAS engine consists of a diesel oxidation catalyst (DOC), ammonia dozer, a selective catalytic reduction catalyst coated on a particulate filter (SCRF), and two underfloor SCR catalysts.

10.4 AIR-HANDLING ARCHITECTURE

The air-handling configuration is illustrated in Fig. 10.4. A dual loop configuration is used to optimize emissions and fuel economy trade-offs. A low pressure EGR (LP EGR) loop draws filtered exhaust gas from downstream of the turbocharger and DOC/SCRF, through a cooler and valve, and mixes it with the fresh air flow upstream of the compressor. The high pressure EGR (HP EGR) loop draws exhaust from upstream of the turbocharger and mixes it with the fresh air/LP EGR mixture downstream of the compressor and charge air cooler. In addition to the LP and HP EGR valves in the system, there is also an ET downstream of the SCRF (which is used to create back pressure to drive the LP EGR flow), and a VGT. Effectively using all four of these actuators to meet the air-handling requirements is the key control problem to be addressed.

TABLE 10.1 Engine Architecture and Specifications

Engine displacement	2800 cc
Cylinder configuration	Inline-4
Peak power	210 hp @ 3600 RPM
Peak torque	385 ft-lbs @ 2000 RPM
Stroke	100 mm
Bore	94 mm
Connecting rod length	158 mm
Compression ratio	15.3:1
Number of valves	4
Air-handling configuration	Dual Loop EGR − High Pressure + Low Pressure Cooled
Turbocharger	Variable geometry
Fuel system	Bosch Piezo-2000 bar
Injectors [holes × SA × flow]	8 × 146 × 410 cc/30 s

FIGURE 10.3   ALTAS engine.

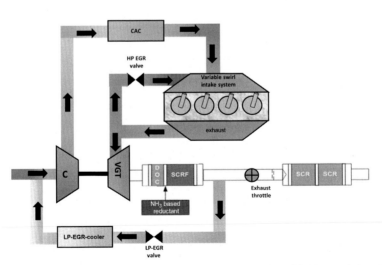

FIGURE 10.4   ALTAS engine full architecture diagram. *Reprinted with permission SAE Copyright © 2014 SAE International. Further distribution of this material is not permitted without prior permission from SAE.*

10.5 TORQUE CURVE AND DUTY CYCLE

The torque curve and operating envelope for the LA-4 cycle is shown in Fig. 10.5. An eight-speed transmission is coupled with this engine in the vehicle, resulting in a narrow operating range of engine speed. Operation within this duty cycle was the focus of the initial air-handling controller development since it constituted a good transient route to challenge the tracking performance of the controller.

10.6 SYSTEM IDENTIFICATION

Unlike the previous chapters which relied solely on the MATLAB *MPC Designer App*, in this chapter we use specialized industrial software, OnRamp Design Suite [1], that was developed by Honeywell. The software streamlines the design of MPC for air-handling control in addition to generating the nonlinear model for the air-handling system. OnRamp Design Suite still relies on MATLAB and Simulink environment to execute, but it has its own MPC toolbox that is optimized to control air-handling systems. OnRamp also allows the generation of a nonlinear model for the air-handling system. It also enables easier deployment of the developed MPC with the engine Electronic Control Module. One difference between OnRamp and the process that we highlight in this book is the way the linear models are generated. In OnRamp, after the input/output data is collected from the engine testing, a nonlinear model for the engine and air-handling system is automatically generated. The linear models are then derived by OnRamp from the nonlinear model (Fig. 10.6).

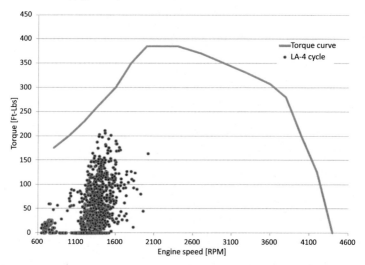

FIGURE 10.5 The engine torque curve, and operating envelope for the LA-4 duty cycle. *Reprinted with permission SAE Copyright © 2014 SAE International. Further distribution of this material is not permitted without prior permission from SAE.*

To calibrate the nonlinear air-handling model, a design of experiment (DOE) was run on the engine in which each of the actuators is swept at a variety of speed fuel points. For the ATLAS engine, the DOE required 4 days of test cell measurement time.

The data obtained from the DOE campaign were then used in an automatic model identification process (described in detail in [17]). The model calibration process consisted of three-steps:

- Identification of individual components: Each of the components (e.g., valves, heat exchangers, turbochargers, etc.) is fit to its respective input-output data. The identified parameters are then used as a starting point for the second step.
- Steady-state system level identification: This step is used to improve the model accuracy after all the calibrated components are assembled together. This step ensures a good match between the system-level model input/output variables and the measured data that is essential for control design. The steady-state accuracy of the control oriented model for the ATLAS engine is shown in Fig. 10.7.
- Transient system-level identification: Control design also requires a representation of the system level dynamics. This step calibrates the dynamic behavior of the model to match the measured engine data.

The calibrated control-oriented model is used to design the controller as described in the next section.

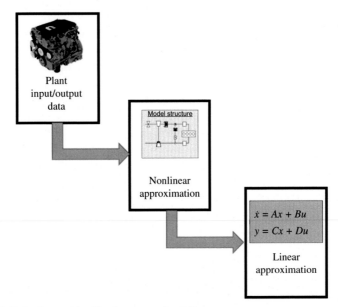

FIGURE 10.6 System identification process for OnRamp.

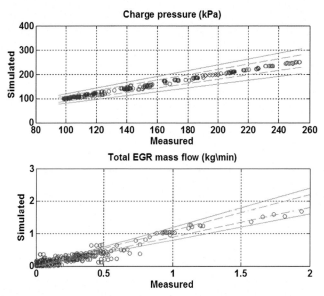

FIGURE 10.7 Steady-state accuracy of the nonlinear model with 10% and 20% accuracy intervals. *Reprinted with permission SAE Copyright © 2014 SAE International. Further distribution of this material is not permitted without prior permission from SAE.*

10.7 MPC CONTROLLER STRUCTURE

The multivariable controller implemented in this work consists of two main parts—feedforward and feedback (Fig. 10.8).

The role of feedforward control is to provide a fast responding, approximate actuator position to the controller during heavy transients. The feedforward part of the control is realized by a set of static lookup tables which are computed through an optimization-based routine performed on the control oriented model of the engine, and the desired setpoints for the control. The output of the lookup tables is filtered by first order filters. The time constants of the filters can be used as additional tuning parameters. The algorithm to compute the feedforward lookup tables is based on a numerical inversion of the control-oriented model in a specified grid of operating points. The tables are parameterized by a selected set of important external variables—in this case, engine speed and fuel injection quantity. The optimization in one point of the grid can be written as:

$$u_{FF}^* = \arg \min_{u, \, \varepsilon} \left\{ \begin{array}{c} \displaystyle\sum_{k \in \text{tracked}} q_k \left(y_k - y_k^{sp}\right)^2 + \displaystyle\sum_{k \in \text{manipulated}} r_k \left(u_k - u_k^{sp}\right)^2 + \\[2ex] \displaystyle\sum_{k \in \text{limited}} w_k \left(y_k - \bar{y}_k - \varepsilon_k\right)^2 + \displaystyle\sum_{k \in \text{limited}} w_k \left(\underline{y}_k - \varepsilon_k - y_k\right)^2 \end{array} \right\}$$

S.t. $\quad \underline{u}_k \le u_k \le \bar{u}_k, \; \forall k \in \text{manipulated}$

$\quad\quad \underline{y}_k - \varepsilon_k \le y_k \le \bar{y}_k + \varepsilon_k, \; \forall k \in \text{limited}$

$\quad\quad \varepsilon_k \ge 0, \; \forall k \in \text{limited}$

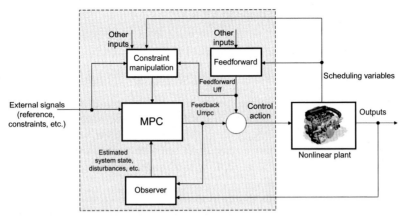

FIGURE 10.8 Controller structure. *Reprinted with permission SAE Copyright © 2014 SAE International. Further distribution of this material is not permitted without prior permission from SAE.*

where q_k, w_k, r_k are weighting (tuning) parameters. For a given point of grid, y_k is k-th output of the model (it is a function of inputs u_k) and y_k^{sp} is the set-point value. $\bar{y}_k$ and $\underline{y}_k$ are the upper and lower limits of the k-th output. u_k^{sp} is the preferred position of k-th actuator, $\bar{u}_k^{sp}$ and $\underline{u}_k^{sp}$ are the upper and lower limits of the k-th manipulated variable. ε_k is the auxiliary variable that enables softening the constraints in case of conflicting or unfeasible limits.

The feedback portion takes care of disturbance rejection, model-engine mismatch (whether from production dispersion or aging), offset-free steady state tracking, and constraints handling. The feedback consists of a MPC controller in which only time-varying limits on actuators were employed, i.e.:

$$\underline{u}(t + k) \leq u(t + k|k) \leq \bar{u}(t + k)$$

The MPC cost function used in this work can be written as:

$$J(u, x(t)) = \sum_{k=0}^{N} \left\| y(t+k|t) - y_{sp}(t) \right\|_Q^2 + \sum_{k=1}^{N_c} \left\| \Delta u(t+k|k) \right\|_R^2$$

where the first term in the cost function is penalization of tracking error over the prediction horizon N, and the second term is penalization of actuator movements over the control horizon Nc, i.e.:

$$\Delta u(t + k|k) = u(t + k|k) - u(t + k - 1)$$

For engine applications where all the computations must be done on an ECM, a fast and reliable QP solver is the key for successful implementation of a MPC in engine control applications. Currently, there are various solvers that can meet these strict requirements (see, e.g., [16]). The QP solver employed in this work is an explicit solver from the Honeywell OnRAMP software suite that has been designed with the goal of meeting the computational speed and ECM footprint requirements.

10.8 CONTROLLER DEPLOYMENT

The MPC controller was designed to control charge pressure and total EGR mass flow. These variables were selected due to their close correspondence to the performance of the system.

The charge pressure was measured directly with a combination manifold pressure and temperature sensor located in the intake manifold. The HP EGR was estimated via subtraction from a charge flow estimator via a speed-density calculation and fresh air flow via a mass air flow sensor in the air box. The LP EGR flow rate was calculated via a delta pressure measurement across a flow orifice. The total EGR was then calculated as the sum of the HP and LP EGR flow rates.

The MPC controller was designed to command four actuators—high and low pressure EGR valves, VGT, and the ET. Since only two variables were tracked, the additional degree of freedom in the controller was addressed by penalizing the movement of the actuators from their feedforward references.

The controlled system is highly nonlinear and therefore it is necessary to use multiple linear MPCs. The MPC controller in this work is structured as a set of local linear MPC controllers that are scheduled (switched on/off)—in this case by engine speed and fuel injection quantity. Note that the local linear models used in MPC controllers were derived from the nonlinear model. The set of operating points has been selected as a regular grid covering the expected operating space of the engine. The total number of linear models was 36, covering engine speed from 800 to 2600 rpm and fuel injection quantity in the range from 0 to 60 mg/stroke. The gain scheduling strategy employed in this work is described in Stewart et al. [18]. Note that additional compensation for the nonlinearities has been applied by using piecewise affine functions at each controlled input and output of the system. The functions have been computed by formulating and solving an optimization problem with the objective to equalize the local linear sensitivities across the whole considered operating space of the engine. Three tuning flavors of the MPC were tested. The tuning strategy in OnRamp is driven by two important parameters—robustness and performance. The robustness of each local linear MPC controller is ensured by using a small gain theorem applied to a user specified level of model uncertainty. The prediction horizon was computed automatically based on the settling time of the linearized dynamic response of the system. The prediction horizon has been in the range from 1 to 5 seconds, depending on the operating point. The controller bandwidth was specified for each actuator independently, by manipulating the corresponding weighting parameters in the MPC cost function.

The developed controller was embedded in a dSpace MicroAutoBox 1401. The MicroAutoBox communicates with the ECM of the engine via a CAN line (Fig. 10.9). The MicroAutoBox receives feedback of the outputs from the ECM in addition to the position of the actuators (VGT position, LP and HP EGR valves, and the ET position). In the MicroAutoBox, the MPC

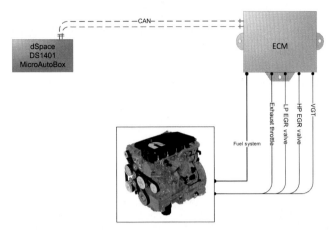

FIGURE 10.9 Hardware setup. *Reprinted with permission SAE Copyright © 2014 SAE International. Further distribution of this material is not permitted without prior permission from SAE.*

computes the tracking errors, evaluates and optimizes for cost function and commands the final position of the actuators which are transmitted via CAN to the ECM. The ECM would then drive the actuators to the commanded positions. The step size for the controller was 0.1 seconds.

10.9   EXPERIMENTAL RESULTS

To test the transient performance of the MPC controller for a light duty vehicle application, it is customary to use the EPA Urban Dynamometer Driving Schedule, which is commonly referred to as the "LA-4" or "the city test," and represents city driving conditions. Before testing the controller in a full transient cycle, it is good practice to test the controller near the 36 operating modes of the controller. Fig. 10.10 shows the tracking performance of the controller at an engine speed of 1800 rpm and total injection quantity of 52.5 mg/stroke when the charge pressure and total EGR mass flow setpoints were step-changed. The dashed lines in Fig. 10.11 show the respective actuators' measured positions. The solid line represents the upper and lower permissible limits of the actuators used as constraints in the MPC formulation. These limits are used as a final safe measure to limit the feedback authority of the actuators during initial testing. These limits were manually relaxed and fixed after initial tuning showed promising results.

After checking the performance of the MPC controller near each modal point, we tested the performance in an LA-4 cycle. To prove the tuning flexibility of the designed MPC controller, we implemented three flavors of the same MPC controller:

- aggressive EGR mass flow tracking
- balanced tuning
- aggressive charge pressure tracking

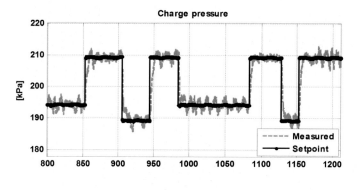

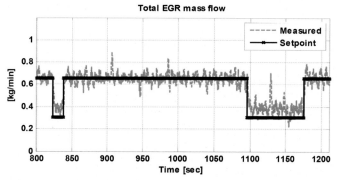

FIGURE 10.10 Tracking performance of the MPC controller at a fixed speed of 1800 rpm and total injection quantity of 52.5 mg/stroke. *Reprinted with permission SAE Copyright © 2014 SAE International. Further distribution of this material is not permitted without prior permission from SAE.*

These three flavors were generated by changing the weights of the cost function of the designed MPC controller. Fig. 10.12 shows the root mean squared error (RMSE) of the controlled variables for the different tunings.

Fig. 10.13 shows the difference in the tracking performance in the time domain for two tunings of the MPC controller. The two figures on the left show the performance of the MPC tuning that is tailored to aggressively track charge pressure whereas the ones on the right show the performance of the tuning that is tailored to aggressively track total EGR mass flow. The total EGR mass flow tracking plot shows a better tracking performance for the MPC controller that was tuned to aggressively track EGR mass flow. Comparing the lower plots of Fig. 10.13 in the time interval [265 290] seconds, one can see the tracking difference between the two flavors of the MPC in tracking the EGR mass flow. Similarly, comparing the top plots of Fig. 10.13, it is apparent that the MPC controller that had an aggressive tuning to track charge pressure outperforms the one that was designed to aggressively track EGR mass flow. Similarly, comparing the upper plots of Fig. 10.13 in the time interval [300 320] seconds, one can see the tracking difference between the two flavors of the MPC in tracking the charge pressure.

Fig. 10.14 shows that as the MPC controller was tuned more aggressively to track EGR mass flow, NOx production was reduced at the expense of

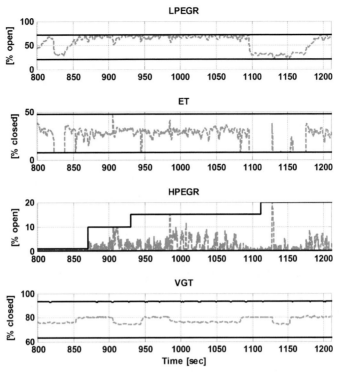

FIGURE 10.11 Actuators' positions corresponding to Fig. 10.10. *Reprinted with permission SAE Copyright © 2014 SAE International. Further distribution of this material is not permitted without prior permission from SAE.*

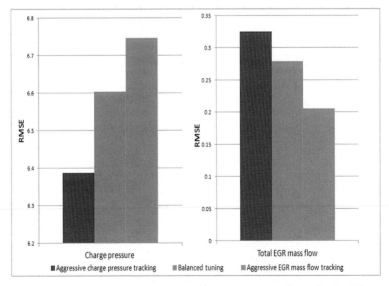

FIGURE 10.12 Root mean squared error of measured controlled variables for the different tunings of the MPC controller. *Reprinted with permission SAE Copyright © 2014 SAE International. Further distribution of this material is not permitted without prior permission from SAE.*

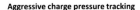

Aggressive charge pressure tracking

Aggressive total EGR mass flow tracking

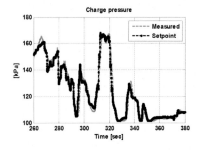

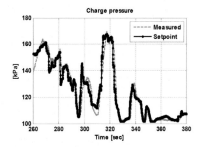

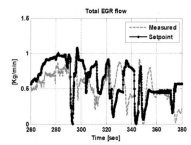

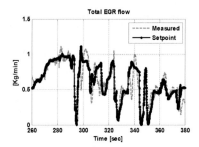

FIGURE 10.13 Root mean squared error of measured controlled variables for the different tunings of the MPC controller. *Reprinted with permission SAE Copyright © 2014 SAE International. Further distribution of this material is not permitted without prior permission from SAE.*

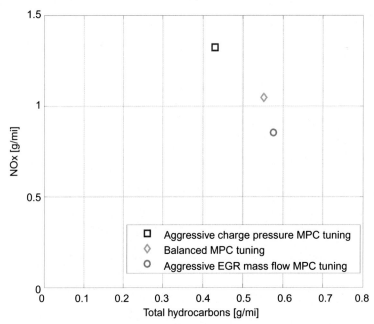

FIGURE 10.14 NOx versus Hydrocarbon tradeoff due to MPC tuning. *Reprinted with permission SAE Copyright © 2014 SAE International. Further distribution of this material is not permitted without prior permission from SAE.*

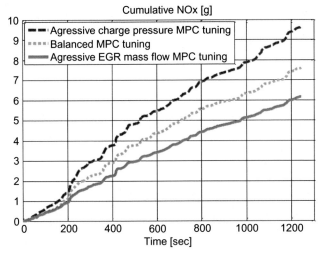

FIGURE 10.15 NOx emissions during the drive cycle for the different MPC tuning options. *Reprinted with permission SAE Copyright © 2014 SAE International. Further distribution of this material is not permitted without prior permission from SAE.*

increased hydrocarbons. Having a direct relation between controller tuning parameters and emissions would enable the tuning team to hit desired emission limits by manipulating the weights of MPC cost functions. This capability is greatly appreciated in a production-intent design process of the air-handling controller where the controller performance is assessed based on emissions and not on tracking performance. Having control of such trade-offs allows the system integrators a margin of flexibility when assessing what type of aftertreatment is needed to meet emissions (e.g., sizing the diesel oxidation and selective reduction catalysts). Fig. 10.15 shows the cumulative NOx for the three controller tunings over the complete drive cycle.

10.10 ROBUSTNESS ANALYSIS

To assess the robustness of the proposed controller against modeling uncertainty, the synthesized MPC controller that was tuned to aggressively track EGR mass flow was combined with the nonlinear model that was developed and calibrated in Section 10.6. The plant + controller model was used to run the Monte-Carlo simulations (similar to Chapter 7). Two thousand LA-4 cycle simulations were run while randomly varying six parameters to simulate the plant uncertainties. The parameters that were varied were: VGT, HP and LP EGR valves, and ET feedforward values. The feedforward values were altered by adding a constant offset per each LA-4 cycle to simulate inadequate feedforward values that were used in the design of the MPC. Furthermore, to simulate the change of the back pressure in the system, the effective area of the DPF was randomly changed as well. Finally, the charge air cooler effectiveness was randomly varied for each LA-4 cycle to simulate

TABLE 10.2 Parameters That Were Used in the Monte-Carlo Simulations

Parameter	Range	Standard Deviation	Mean	Histogram
VGT Command Offset [%]	[−5,5]	5	0	
HP EGR Valve Command Offset [%]	[−5,5]	5	0	
LP EGR Valve Command Offset [%]	[−5,5]	5	0	
ET Command Offset [%]	[−5,5]	5	0	
DPF Effective Area []	[0.5,1]	0.5	1	
Charge Air Cooler Efficiency []	[0.5,1]	0.5	1	

the aging of the system. Table 10.2 shows the values used in generating the random uncertainties. The uncertainties were generated using MATLAB by using the normal distribution and then restricting the uncertainties to the range specified in Table 10.2.

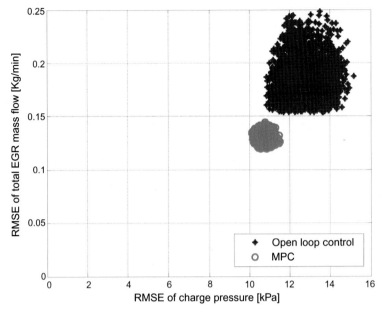

FIGURE 10.16 Performance of MPC and open loop control under the impact of uncertainties. *Reprinted with permission SAE Copyright © 2014 SAE International. Further distribution of this material is not permitted without prior permission from SAE.*

Fig. 10.16 shows the tracking capabilities of the MPC and open loop control when subjected to the same levels of uncertainties. In the open loop control, the feedforward lookup tables that were used in synthesizing the MPC were used without any corrective feedback action. In Fig. 10.16, the RMSE of total EGR mass flow is plotted versus charge pressure for the 2000 simulations. The MPC was successful in reducing the standard deviation of the RMSE of total EGR mass by 73.7% and the standard deviation of the RMSE of the charge pressure by 66.7%. Furthermore, Fig. 10.16 clearly shows the enhanced tracking capability of the closed loop control (MPC) when compared to the open loop control.

10.11 SUMMARY

In this chapter, we applied Honeywell automotive software, OnRamp, to design a MPC for controlling dual loop EGR air-handling architectures. The designed controller was used to handle the complexity of coordinating the various actuators of the air-handling system. The process of generating the MPC controller was validated experimentally on a light duty diesel engine. The tracking performance of the controller was linked to emissions measures through the tuning parameters of MPC. Robustness of the synthesized MPC was validated through Monte-Carlo simulations.

The duration of time it took to develop and experimentally test the MPC controller was about 10 days. Given the degree of complexity of the architecture and the duration it took to design the MPC and test the closed-loop performance of the system, such a control technique provides a fast process to assess the capability of various proposed engine architectures.

Industrial applications for MPCs are expected to increase with the increased demand for smart devices. Industrial software such as OnRamp that facilitate MPC design and deployment, will become more popular during the next decade.

REFERENCES

[1] Honeywell OnRAMP website: http://www.honeywellonramp.com
[2] G.E. Stewart, F. Borrelli, J. Pekar, D. Germann, D. Pachner, D. Kihas, Toward a systematic design for turbocharged engine control, in: L. del Re, F. Allgower, L. Glielmo, C. Guardiola (Eds.), "Automotive Model Predictive Control", Lecture Notes in Control and Information Science, Springer-Verlag, Berlin-Heidelberg, 2010.
[3] N. Khaled, J. Pekar, A. Fuxman, M. Cunningham, O. Santin. Multivariable control of dual loop EGR diesel engine with a variable geometry turbo. 2014-01-1357, SAE World Congress Conference, Detroit, 2014.
[4] A. Suresh, D. Langenderfer, C. Arnett, M. Ruth, Thermodynamic systems for Tier 2 Bin 2 diesel engines, SAE Int. J. Eng. 6 (1) (2013) 167−183. Available from: https://doi.org/10.4271/2013-01-0282.
[5] J. Shutty, R. Czarnowski. Control strategy for a dual loop EGR system to meet Euro 6 and Beyond. Deer Conference, August 3−6, Dearborn, Michigan, 2009.
[6] F. Millo, P.F. Giacominetto, M.G. Bernardi, Analysis of different exhaust gas recirculation architectures for passenger car Diesel engines, Appl. Energy 98 (1) (2012) 79−91.
[7] M. van Aken, F. Willems, D. de Jong. Appliance of high EGR rates with a short and long route EGR system on a heavy duty diesel engine. SAE Technical Paper 2007-01-0906, 2007, https://doi.org/10.4271/2007-01-0906.
[8] M. Zheng, G.T. Reader, J.G. Hawley, Diesel engine exhaust gas recirculation-a review on advanced and novel concepts, Energy Conv. Manage. 45 (1) (2004) 883−900.
[9] J. Shutty, H. Benali, L. Daeubler, M. Traver. Air system control for advanced diesel engines. SAE Technical Paper 2007-01-0970, 2007, https://doi.org/10.4271/2007-01-0970.
[10] O. Grondin, P. Moulin, J. Chauvin. Control of a turbocharged Diesel engine fitted with high pressure and low pressure exhaust gas recirculation systems. Proceedings of the 48th IEEE Conference on Decision and Control held jointly with the 28th Chinese Control Conference, 6582−6589, 2009, https://doi.org/10.1109/CDC.2009.5400922.
[11] F. Yan, J. Wang. Control of dual loop EGR air-path systems for advanced combustion diesel engines by a singular perturbation methodology. Proceedings of the American Control Conference, 1561−1566, 2011.
[12] J. Shutty, R. Czarnowski. Control strategy for a dual loop EGR system to meet Euro 6 and beyond. Presented at DEER Conference, USA, August 3−6, 2009.
[13] B. Haber, "A Robust Control Approach on Diesel Engines with Dual-Loop Exhaust Gas Recirculation Systems," Master Thesis, The Ohio State University, 2010.
[14] J. Wang, Air fraction estimation for multiple combustion mode diesel engines with dual-loop EGR systems, Control Eng. Practice 16 (1) (2008) 1479 1486.

[15] A. Bemporad, M. Morari, V. Dua, E.N. Pistikopoulos, The explicit linear quadratic regulator for constrained systems, Automatica 38 (2002) 3−20.

[16] F. Borelli, M. Baotic, J. Pekar, G. Stewart, On the computation of linear model predictive control laws, Automatica 46 (2010) 1035−1041.

[17] D. Pachner, D. Germann, G.E. Stewart, Identification techniques for control oriented models of internal combustion engines, in: D. Alberer, H. Hjalmarsson, L. del Re (Eds.), "Identification for Automotive Systems", Lecture Notes in Control and Information Science, Springer-Verlag, Berlin-Heidelberg, 2012.

[18] G.E. Stewart, F. Borrelli. System for gain scheduling control. USPTO number 7603185, October 2009.

Index

Note: Page numbers followed by "*f*" and "*t*" refer to figures and tables, respectively.